The Urban Climate

The Urban Climate

HELMUT E. LANDSBERG

Institute for Physical Science and Technology
University of Maryland
College Park, Maryland

1981

ACADEMIC PRESS

A Subsidiary of Harcourt Brace Jovanovich, Publishers

New York   London   Toronto   Sydney   San Francisco

COPYRIGHT © 1981, BY ACADEMIC PRESS, INC.
ALL RIGHTS RESERVED.
NO PART OF THIS PUBLICATION MAY BE REPRODUCED OR
TRANSMITTED IN ANY FORM OR BY ANY MEANS, ELECTRONIC
OR MECHANICAL, INCLUDING PHOTOCOPY, RECORDING, OR ANY
INFORMATION STORAGE AND RETRIEVAL SYSTEM, WITHOUT
PERMISSION IN WRITING FROM THE PUBLISHER.

ACADEMIC PRESS, INC.
111 Fifth Avenue, New York, New York 10003

United Kingdom Edition published by
ACADEMIC PRESS, INC. (LONDON) LTD.
24/28 Oval Road, London NW1   7DX

Library of Congress Cataloging in Publication Data

Landsberg, Helmut Erich, Date.
   The urban climate.

   (International geophysics series; v. 28)
   Includes bibliographies and index.
   1. Urban climatology. I. Title. II. Series.
QC981.7.U7L36   551.5'09173'2   80-2766
ISBN 0-12-435960-4   AACR2

PRINTED IN THE UNITED STATES OF AMERICA

83 84   9 8 7 6 5 4 3 2

Contents

v

4 Urban Energy Fluxes

5 The Urban Heat Island

6 The Urban Wind Field

7 Models of Urban Temperature and Wind Fields

8   Moisture, Clouds, and Hydrometeors

9 Urban Hydrology

10 Special Aspects of Urban Climate

11 Urban Planning

Preface

A quarter century has elapsed since the last monographic review of knowledge about urban climates appeared in book form.[1] That summarization came at a time when a certain plateau of knowledge had been reached. Essentially, it marked the end of a descriptive, geographic phase that established beyond doubt the fact that urban areas have climates different from their surroundings.

Since that time emphasis has been on analytical studies directed toward physical understanding of the rural–urban differences in the atmospheric boundary layer. Attempts to quantify conditions have met with some success. There is certainly a clear understanding of the physical relations that create the climatic differences of urbanized areas. This understanding represents a new plateau of knowledge, deserving summarization, and that is the aim of this book.

No attempt has been made here to give comprehensive coverage of the literature. Although some of the earlier classical studies are cited, the emphasis is on the work done during the last decade and a half.

It is my hope that the text will not only be useful for boundary layer meteorologists, but also for city planners and developers as well as human ecologists.

[1] Kratzer (1956); see p. 15, this volume, for complete reference.

I am very much obliged to my secretary, Mrs. Katherine Mesz-tenyi, who had to suffer through drafts and revisions in typing the manuscript. For excellent graphics support I am grateful to Ms. Clare Villanti. My gratitude also goes to the National Science Foundation, which from 1968 to 1974 supported surveillance of climate during the development of the new town of Columbia, Maryland.

1

Introduction

1.1 THE LITERATURE

The field of urban climatology has grown rapidly in recent years. This is reflected in the phenomenal increase in literature. When Father Kratzer wrote his first book on the subject, an outgrowth of his dissertation, he cited 225 papers (Kratzer, 1937). The second edition (Kratzer, 1956) listed 533 publications. A selective, annotated bibliography by C. E. P. Brooks (1952) found 249 items worthy of abstracting. That was about the time when the modern era of investigation started. Thus we find in Chandler's (1970) comprehensive bibliography, prepared for the World Meteorological Organization, some 1800 titles. This was followed by a review of progress between 1968 and 1973 by Oke (1974) citing 377 papers that had appeared since Chandler's work. The same author again reviewed the increment in pertinent literature between 1973 and 1976 (Oke, 1979) and found 434 references worthy of citation. A recent bibliography restricted to Australian contributions gives some 554 titles, some of only peripheral relation to the urban climate theme.

This enormous increase in literature reflects the growing concern about man's influence on his environment. This concern kindled all kinds of inquiries in many parts of the world. The efforts to gain insight into the man-made climatic alteration in cities ranged from the profound to the trivial. The momentum gathered in recent years will carry these inquiries forward for a while yet. Such work can elucidate further details here and there or add other towns to the list for which data are obtained, generally only confirming older results. The main remaining tasks are to translate these results for use by urban planners and to clarify that elusive mechanism of urban-induced precipitation and its areal extent.

A major effort was made to come to grips with the latter problem in project METROMEX, which was a comprehensive study of the atmospheric effects of a major metropolitan area. Extensive reports have been rendered (Changnon *et al.*, 1977; Ackerman *et al.*, 1978), and a whole issue of the *Journal of Applied Meteorology* has been devoted to general publication of results of this project (Changnon, 1978). Any student interested in the topic of urban climatology must refer to these original studies. It was this project that brought this field of scientific inquiry to its present plateau. Because of the vastness of the literature (well in excess of 3000 titles), its recent repetitive character, and the excellent bibliographical resources cited earlier, only direct sources of material given in this book will be cited henceforth.

We may be justified in asking: Why has the topic reached such dimensions? The reason is simple. Cities and metropolitan areas have grown into vast conurbations, where a variety of human influences impinge on the local atmosphere—and the end is not in sight. United Nations projections indicate that by the end of the twentieth century over 6 billion people will live on earth, about 50 percent of them crowded into urban areas. In the so-called developed world the expectation is for about 1.4 billion inhabitants with about 80 percent living in cities (Fig. 1.1). In a few short years some metropolitan areas will reach the 20-million population figure. Clearly, in addition to the vast changes that have already taken place, further transformations of land must take place to accommodate these people. More fuel must be used, more waste heat dissipated, more pollutants dispersed to furnish housing, employment, and transportation to these masses. All of this impinges on the atmosphere. The ecological impact is enormous. This has been a topic for debate for over a decade. A number of these discussions have found their way into print and the reader should take time to look at the larger context. Urban climate is just one small facet of a much larger problem facing mankind (see, e.g., Eldridge, 1967; Dansereau, 1970).

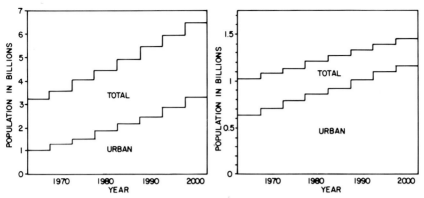

Fig. 1.1 United Nations projections of population in the world (left) and in developed areas (right) to the year 2000, with the percentage living in urban areas.

1.2 HISTORICAL DEVELOPMENTS

Ever since cities developed in antiquity, people noticed that urban air was different from rural air. They sensed a persistent evil of cities with that highly sensitive chemical monitor, the nose. The reaction is to air pollution. Although the sources have changed through the ages, polluted air seems to be the hallmark of the urban atmosphere.

An allusion to Roman smoke pollution already appears in the odes of Quintus Horatius Flaccus (Horace, 65–68 B.C.) about 24 B.C. (Neumann, 1979). More specifically we read in the writings of Lucius Annaeus Seneca (ca. 3 B.C.–A.D. 65): "As soon as I had left the heavy air of Rome with its stench from smoky chimneys which, when stoked, will belch their enclosed pestilential vapor and soot, I felt a change in mood."

In the outgoing Middle Ages London became the prototype of urban pollution. It prompted prohibition of coal burning in 1273. This was to little avail. In 1306 King Edward I (1239–1307) issued a proclamation banning the burning of sea coal in furnaces. Queen Elizabeth I (1533–1603) banned the burning of coal in the city during sessions of Parliament. The great naturalist, John Evelyn (1620–1706), a Fellow of the Royal Society, wrote a pamphlet in 1661 decrying the use of coal for manufacturing. In it he said: "For in all other places the Aer is most Serene and Pure, it is here Ecclipsed with such Cloud of Sulphure as the Sun itself, which gives day to all the World besides, is hardly able to penetrate and impart it here; and the weary Traveller at many Miles distance, sooner smells, than sees the City to which he repairs." The pollution problem persisted for nearly another 300 years until an air pollution episode in 1952 caused 4000 excess deaths and led to legislation that converted this and other English cities into "smokeless" zones.

London appeared again in the annals of urban climatology when in 1818 Luke Howard (1772–1864, Fig. 1.2) published the first edition of a book dealing with the climate of the city. A second volume appeared in 1820 and a third, enlarged edition was published in 1833. Howard, a chemist, was a pioneering amateur meteorologist. His cloud classification of 1803 is still the basis for cloud identification. In his book, also incidentally the first monographic treatment of the climate of a city, Howard clearly recognized a major alteration of a meteorological element. For this he created the term "city fog." He

Fig. 1.2 Portrait of Luke Howard (1772–1864), FRS, discoverer of London's heat island.

vividly described the phenomenon for several cases. About January 10, 1812 he wrote:

> London was this day involved, for several hours, in palpable darkness. The shops, offices &c were necessarily lighted up; but the streets not being lighted as at night, it required no small care in the passenger to find his way and avoid accidents. The sky, where any light pervaded it, showed the aspect of bronze. Such is, occasionally, the effect of the accumulation of smoke between two opposite gentle currents, or by means of a misty calm. I am informed that the fuliginous cloud was visible, in this instance, for a distance of forty miles. Were it not for the extreme mobility of the atmosphere, this volcano of a thousand mouths would, in winter be scarcely habitable.

The colorful description of the urban scene as a volcano has been repeated in many writings since. The strict localization of the area of low visibility or fog was described by Howard for the case of January 16, 1826 as follows: "At one o'clock yesterday afternoon the fog in the city was as dense as we ever recollect to have known it. Lamps and candles were lighted in all shops and offices, and the carriages in the street dared not exceed a foot pace. At the same time, five miles from town the atmosphere was clear and unclouded with a brilliant sun."

However, the most remarkable discovery of Howard was his recognition that the urban center was warmer than the surrounding countryside. We reproduce here a table published by Howard from a comparison of thermometer readings in London and in the country. This appeared in 1820. In it he shows the mean temperature difference for each month (Fig. 1.3). Particularly noteworthy is the sentence in the footnote to the table: "Night is 3.70° *warmer* and day 0.34° *cooler* in the city than in the country." This in a nutshell is the recognition of the urban heat island, a topic that will weave through the remainder of this book. Howard, as many after him, attributed the greater temperature of the city to the extensive use of fuel.

From another metropolis, Paris, new evidence for an altered climate was published a few decades later by E. Renou (1815–1902, Fig. 1.4). In 1855, in connection with a set of instructions on meteorological observations, Renou was concerned about exposure of thermometers and the reality of the apparent temperature increment in the city. In commenting about observations indicating 1–2°(C) he says:

> It cannot be otherwise; the interiors of houses in Paris have annual means, varying from 12 to 16°; enormous quantities of water in all forms pour out. The respiration of humans and animals, above all the fumes of innumerable

Results in Figures.

Average and difference of Day and Night for each month.

Mo.	Mean of greatest heat by Day.	Mean of greatest cold by Night.	Difference.
1. Jan.	40·28	31·36	8·92
2. Feb.	44·63	33·70	10·93
3. Mar.	48·08	35·31	12·77
4. April	55·37	39·42	15·95
5. May	64·06	46·54	17·52
6. June	68·36	49·75	18·61
7. July	71·50	53·84	17·66
8. Aug.	71·23	53·94	17·29
9. Sept.	65·66	48·67	16·99
10. Oct.	57·06	43·51	13·55
11. Nov.	47·22	36·49	10·73
12. Dec.	42·66	33·90	8·76

Extremes of the Climate. Greatest heat in 10 years 96°; greatest cold —5 (below zero). Difference of *night* from *day* sometimes 30° or 35°; seldom less than 6°. Night is 3.70 *warmer* and day 0.31° *cooler* in the city than in the country. Thus the latter has 4° more variation.

Mean Temperature of each Month, on an average of observations continued from 1807 to 1816.

Mo.	In the Country.	In London.	London warmer.
1. Jan.	34·16°	36·20°	2·04
2. Feb.	39·78	41·47	1·69
3. Mar.	41·51	42·77	1·26
4. April	46·89	47·69	0·80
5. May	55·79	56·28	0·49
6. June	58·66	59·91	1·25
7. July	62·40	63·41	1·01
8. Aug.	61·35	62·61	1·26
9. Sept.	56·22	58·45	2·13
10. Oct.	50·24	52·23	1·99
11. Nov.	40.93	43·08	2·15
12. Dec.	37·66	39·40	1.74

By this Table, the reader who makes daily observations on the temperature for a month may compare his mean result with a fixed standard.

Fig. 1.3 Luke Howard's table of urban temperatures (°F) and urban–rural temperature differences.

Fig. 1.4 Photograph of Emilien Renou (1815–1902). [Courtesy of Météorologie Nationale (Photo 10.027).]

chimneys, maintain above Paris a rust-colored haze which blocks the sun if one looks to the city from the heights which rise at some distance, for example, the Terrace of Mendon; it is impossible that this turbid, smoky, ammoniacal atmosphere should not have a notably higher temperature than the surrounding country.

We are not exactly sure if Renou was suggesting there radiative energy absorption, industrial heat rejection, or metabolic heat as the cause or causes for the warmer urban air temperature. He returns to this theme over a decade later (Renou, 1868) in a paper that critically compares temperatures at a number of sites inside and outside Paris. He is still quite concerned with the problem of thermometer exposure. His conclusion is that: "clearly the temperature difference between the countryside (and the city) is about 1°(C), at the same elevation."

He further states: "The cities, particularly the large ones, retard the march of temperature and thus alternate the oscillations, especially the most abrupt ones. This is mainly notable for the evening cooling in clear, calm weather chiefly when there is a touch of fog, which veils the cities more readily than the countryside. . . ." Here the rather astute observation is made that the evening differences between city and surroundings are greatest when the weather favors strong radiative cooling. He also notes the considerable difference of days with freezing temperatures in the two environments, with the comparable rural station having about 40 percent more days with freezing temperature in a test year. He finally noted that wind speeds in town are lower than in the countryside.[1]

In the next half-century a number of monographic studies of climates of various cities appeared. In most of these are comments on the urban–rural climatic differences and a few short papers specifically relating to such differences were published. However, systematic research relating to both the phenomenology and the causes of these differences did not become prominent until Wilhelm Schmidt (1883–1936, Fig. 1.5), started micrometeorological investigations of the urban landscape (Schmidt, 1917). One of his major contributions was the introduction of the use of instrumented motor vehicles as a research tool (Schmidt, 1927, 1930). This permitted the plotting of isolines of meteorological elements, essentially on a synoptic basis, with time corrections made by use of automatic records at one or more fixed points. Since that time instruments on other vehicles, including bicycles and streetcars, have served such survey purposes. Nowadays, helicopters and satellites have been added to the arsenal.

It should also be recalled that August Schmauss (1877–1954) asserted an atmospheric influence of the city of Munich, causing rain increases toward the lee (Schmauss, 1927). There were some earlier studies on rainfall trends in large urban areas that suggested increases as the cities grew.

The rapid increase in the size of metropolitan areas after World War II and the rising industrialization, with alarming changes in aerosols, deterioriation of visibility, and many other manifest atmospheric alterations, led to a multiplicity of studies of urban climates.

---

[1] I am indebted to Mr. J. Dettwiller of the Météorologie Nationale for sending me copies of Renou's papers.

Fig. 1.5 Photograph of Wilhelm Schmidt (1883–1936). (Courtesy of Zentralanstalt für Meteorologie und Geodynamik, Wien.)

Their aim was not only the desire to increase knowledge but also to abate the nuisances, if not acute dangers. Also, the rebuilding of war-damaged cities and the necessary urban renewal led to a growing interest in incorporating climatic as well as other factors of the physical environment into urban planning.

1.3 THE PROBLEM OF LOCAL CLIMATE MODIFICATION

From the foregoing it is quite clear that our senses, especially vision and smell, can adequately verify some effects of urbanization on the atmosphere. All science can do is to put numbers on the modifications that have taken place. In the case of conventional meteorological elements (temperature, humidity, wind, precipitation), it is not as easy to establish the magnitude of modulation of the atmospheric boundary layer by a city. It is even more difficult to assign causes for empirically observed changes.

One of the main difficulties is the geographical setting of many cities. Usually the siting of a city is not haphazard. Settlement has developed for a specific reason. In earlier years riverbeds offered good communications, and that is where we find many towns. In many developed countries sites were selected because they were more readily defensible than others. Coastal cities developed near natural harbors, others near natural resources. In the majority of cases the topography was not simple and there were micro- and mesoclimatic differences between the sites and the surroundings even before the cities ever sprung up. It has been pointed out time and again that the simple comparison, once in vogue, of a weather station in town with one at an outlying airport to establish the urban climatic difference is an inadequate analytical tool. Only in the simplest topographic setting of flat terrain and close proximity can some valid information be collected. However, airports are often located at considerable distances from cities and at sites that offer microclimatic advantages to aviation. The climates at those sites may be *a priori* different from the city locations.

Only if one has observations from a site prior to urbanization, preferably for a considerable length of time, can one make a valid "before and after" comparison. This opportunity is rarely available.

The general dilemma has been discussed in depth by William P. Lowry (1977), part of whose argument we pursue in the following paragraphs.[2] It must be said at the outset that the framework he proposed is idealized and in practice may well be very difficult to realize.

The basic model envisages three basic elements influencing the measured value of a metropolitan variable M. These are: the basic climate of the region C, a difference introduced by location L (i.e., topography, waterbodies, etc.), and an alteration term U produced by urbanization; M will usually be a time series, which constitutes a

---

[2] The symbols used here are somewhat changed from Lowry's presentation.

statistical ensemble. The model is then a sum of the three components

$$M = C + L + U \tag{1.1}$$

Only M is a known quantity. The question is how to estimate the
others. Lowry proposes to derive values for these unknowns by establishing statistical ensembles of the variables for various weather
types. Then, establishing the values of M for a time 0, using a sample
in the beginning of the time series and one at a later time t, and,
assuming invariance of C and L, one can determine U:

$$M_t = C + L + U$$
$$M_0 = C + L + 0 \tag{1.2}$$

In this scheme the urban influence is set to be negligible at $t = 0$, but
one can use the system only in the absence of trends in C, certainly
not a safe assumption for a number of elements. No actual examples
have been worked out in this fashion.

There are other schemes to establish an urban influence, also discussed by Lowry. These have been traditionally used. The simplest
of these is to establish an urban–rural difference of variables measured at one or more sites in the two environments

$$U = M_u - M_r \tag{1.3}$$

where the subscripts u and r stand for urban and rural, respectively.
This is the time-honored approach since the days of Luke Howard.
In simple, flat terrain and in the absence of large waterbodies, it is a
quick approach that can give at least an estimate of the urban effect.
The difference between M_u and M_r is, of course, amenable to the
usual statistical tests of significance.

Another frequently used approach is the comparison of time
trends in the two environments. Assuming that C and L are not time
dependent one can state that

$$\frac{\partial}{\partial t}[M_{t_u} - M_{t_r}] = \frac{\partial}{\partial t}\left[\sum_{i=1}^{n} p_{i_t} M_{t_u}\right] \tag{1.4}$$

where i represents individual values of a time series of n members,
such as successive yearly values, t is time, and p_{i_t} represents the
probability that a given weather type will occur in a given time interval t, perhaps suitably split by seasons.

Under the assumption that the industrial workweek and traffic create differences between workdays and weekends in human influences on some atmospheric parameters, a comparison of Monday–Friday values with Saturday–Sunday values may reveal urban effects:

$$\Delta M = M_w - M_n \qquad (1.5)$$

where M_w is a mean value of the element tested on workdays and M_n the value for nonworkdays. This is not a very powerful test because there may be steady trends from the beginning to the end of the workweek. There are also differences in the total numbers of workdays and nonworkdays. Finally, there is a need to demonstrate that there are no differences in the values for a comparable split of days simultaneously in the same area not affected by urbanization. This last condition is hardest to fulfill in the absence of knowledge of how far from a city its atmospheric influence reaches. Useful results of this type of comparison are almost entirely restricted to measurements of pollution.

For pollutants and certain other elements such as wind speed and perhaps precipitation a profile—upwind rural area, city, downwind rural area—can be revealing. This has to be tested for each wind direction, at least in quadrants. If there are notable urban influences one has to show by suitable statistical tests that

$$M_v \neq M_u \neq M_l \qquad (1.5)$$

where M_v is an upwind mean value, M_u the urban value, and M_l the lee value of the element tested. Here again the topography may seriously interfere or completely vitiate the validity of the test. In flat country, however, this procedure has value.

Two procedures are far more objective, but one of them can rarely be applied. These are physical model tests in the wind tunnel and measurements before, during, and after urbanization. The latter is a technique that is only applicable for new towns and perhaps in large redevelopment projects, where all old structures are removed. In such projects the natural trends or fluctuations of climate have to be removed by using suitable control sites in the area (Landsberg, 1979).

In the wind-tunnel studies a scale model of a city is placed into the airstream of a wind tunnel and measurements are made upstream, in the model city, and downstream. The flow can be made visible by

Fig. 1.6 Wind-tunnel model of urban airflow, simulating Denver, Colorado. (Courtesy of Professor J. E. Cermak, State University of Colorado.)

smoke and can be recorded through moving pictures. A suitable ionizing tracer can also be used. Similar wind-tunnel techniques can be used to determine pressure distributions on tall buildings in an urban setting (Cermak, 1974; SethuRaman and Cermak, 1974). Figure 1.6 vividly shows the turbulence induced over an urban area by a tall structure.

The wind-tunnel studies require similarities in order to simulate natural conditions realistically. Of these conditions only one is easily achieved, namely geometric similarity. This simply means that an exact scale model is placed into the wind tunnel. It is more difficult to achieve thermal similarity of the simulated urban surface and surrounding topography and in the thermal stratification of the air in the wind-tunnel stream. Most difficult to achieve is dynamic similarity. This depends primarily on the Reynolds number[3] of the flow, which when exceeding a certain value indicates turbulent flow. In the wind tunnel both laminar and turbulent flow can be achieved. The laminar flow prevails under stable thermal stratification, i.e., inversion conditions with warm air over a cold surface layer. In nature even under the inversion there is turbulence and hence the dynamic similarity is not perfect. Nonetheless, the wind-tunnel experiments have shown the wind-flow modification by urban areas, the eddy disturbances brought about by tall buildings at street level, and the dispersal characteristics for pollutants released at various points in the flow field.

The procedure to observe the man-induced changes before, during, and after urbanization of a piece of land is straightforward, but the opportunity to apply it arises only rarely. One such occasion offered itself in 1967 when the planned community of Columbia, Maryland in the Baltimore, Maryland–Washington, D. C. corridor began. Use of ordinary, fixed meteorological stations and mobile surface and aerial surveys permitted following the gradual changes in various elements and comparing these with the changes obtained by other means in existing cities (Landsberg, 1979).

References

Ackerman, B., Changnon, S. A., Dzurisin, G., Gatz, D. F. , Grosh, R. C., Hilberg, S. D., Huff, F. A., Mansell, J. W., Ochs, III, H. T., Peden, M. E., Schickedanz, P. T., Semonin, R. G., and Vogel, J. L. (1978). Summary of METROMEX, Vol. 2: Causes of Precipitation Anomalies. *Ill. State Water Surv. Bull.* **63,** 395 pp.

Brooks, C. E. P. (1952). Selective annotated bibliography on urban climate. *Meteorol. Abstr. Bibliog.* **3** (17), 734–773.

Cermak, J. E. (1974). Applications of fluid mechanics to wind engineering—A Freeman Scholar Lecture. *J. Fluids Eng.* **97,** 9–38.

Chandler, T. J. (1970). Selected bibliography on urban climate. *WMO Publ.,* No. 276, T. P. 155, 383 pp.

[3] Reynolds number is Re $= uz/\nu$ where u is the flow (wind speed), z a height parameter, and ν the kinematic viscosity of the air (or fluid).

Changnon, Jr. (ed.) (1978). METROMEX issue. *J. Appl. Meteorol.* **17** (5), 565–715.

Changnon, Jr., S. A., Huff, F. A., Schickedanz, P. T., and Vogel, J. L. (1977). Summary of METROMEX, Vol. 1: Weather Anomalies and Impacts. *Ill. State Water Surv. Bull.* **62**, 260 pp.

Dansereau, P. (ed.) (1970). "Challenge for Survival," 235 pp. Columbia Univ. Press, New York.

Eldridge, H. W. (ed.) (1967). "Taming Megalopolis," Vol. I, 576 pp.; Vol. II, pp. 577–1168. Anchor Books, Doubleday, Garden City, New York.

Evelyn, J. (1661). Fumifugium: Or the Inconvenience of the Air and Smoke of London Dissipated, 43 pp. Oxford. (Reprinted, National Smoke Abatement Society, Manchester, 1933.)

Howard, L. (1833). "Climate of London Deduced from Meteorological Observations," 3rd ed., in 3 Vols. Harvey & Darton, London.

Howard, L. (1851). A companion to the thermometer, for the climate of London. Folio Broadside, reprinting a statement, dated 8 May 1820. Darton & Co., Holborn Hill.

Kratzer, A. (1937). "Das Stadtklima," 143 pp. Friedr. Vieweg & Sohn, Braunschweig.

Kratzer, A. (1956). "Das Stadtklima," 2nd ed., 184 pp. Friedr. Vieweg & Sohn, Braunschweig.

Landsberg, H. E. (1979). Atmospheric changes in a growing community (the Columbia, Maryland experience). *Urban Ecology* **4**, 53–81.

Lowry, W. P. (1977). Empirical estimation of urban effects on climate: A problem analysis. *J. Appl. Meteorol.* **16**, 129–135.

Neumann, J. (1979). Air pollution in Ancient Rome. *Bull. Am. Meteorol. Soc.* **60**, 1097.

Oke, T. R. (1974). Review of urban climatology, 1968–1973. *WMO Publ., Tech.* Note 134, 132 pp.

Oke, T. R. (1979). Review of urban climatology. *WMO Publ.,* Tech. Note 169, 100 pp.

Renou, E. (1855). Instructions météorologiques. *Annuaire Soc. Météorol. de France,* Vol. 3, Part 1, 73–160.

Royal Meteorological Society Australian Branch (1978). Bibliography on urban meteorological studies in Australia, 201 pp. CSIRO, Melbourne.

Schmauss, A. (1927). Groszstädte und Niederschlag. *Meteorolog. Z.* **44**, 339–341.

Schmidt, W. (1917). Zum Einfluss grosser Städte auf das Klima. *Naturwissenschaften* **5**, 494–495.

Schmidt, W. (1927). Die Verteilung der Minimumtemperaturen in der Frostnacht des 12.5.1927 im Gemeindegebiet von Wien. *Fortschr. Landwirtsch.* **2** (H.21), 681–686.

Schmidt, W. (1930). Kleinklimatische Aufnahmen durch Temperaturfahrten. *Meteorolog. Z.* **47**, 92–106.

SethuRaman, S., and Cermak, J. E. (1974). Physical modeling of flow and diffusion over an urban heat island. *In* "Turbulent Diffusion in Environmental Pollution," (F. N. Frenkiel and R. E. Munn, eds.), pp. 223–240. Academic Press, New York.

2

The
Assessment
of the
Urban
Atmosphere

2.1 SYNOPTIC VERSUS LOCAL SCALE

The urban climate cannot be viewed in isolation. It is, like all climates, the statistical composite of the many daily weather events of an area. As such, conditions at any locality are governed by the large-scale weather patterns obvious on a synoptic weather map. Yet each environment modifies more or less the local conditions in that thin air stratum above the ground called the atmospheric boundary layer. Some refer to these imprints of surface conditions as the meso- or microclimates. Because the definitions of these terms are vague and because the scales in the horizontal and vertical dimensions change, the term "local" is preferred here.

The interaction between the synoptic scale and the local scale is a continuous seesaw. Sometimes the large-scale weather conditions are the dominant influences and at others the local conditions are prevalent, although both of them are always present. As a general rule one can state that during a strong synoptic flow pattern, characterized by brisk winds, clouds, and, at times, precipitation, local influences are largely suppressed. However, when winds are weak and the sky is clear, with sunshine during the day and cloudless conditions at night, the local effects control the lowest layer of the atmosphere.

There are interactions between static and dynamic elements and numerous feedbacks. A diagrammatic presentation of the many influences is shown in Fig. 2.1, where arrows indicate which parameter is affected. Some relations are one sided, others work both ways. A few atmospheric variables are influenced by many factors. A good example is visibility. It is influenced by the synoptic-scale value of atmospheric moisture and by the occurrence of precipitation; the pollutant concentration is a local factor. There are also many interactions on the local scale alone. Local vertical and hori-

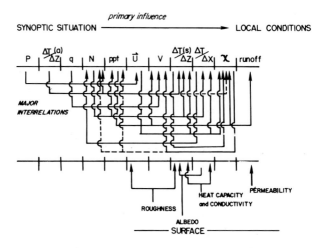

Fig. 2.1 Schematic diagram of interaction between synoptic scale and local-scale parameters:

P	pressure	$\Delta T/\Delta Z(a)$	vertical temperature structure aloft
N	cloudiness		
V	visibility	$\Delta T/\Delta Z(s)$	low-level vertical temperature structure
ppt	precipitation		
U	wind vector		
q	moisture content	$\Delta T/\Delta X$	local horizontal temperature distribution (heat island)
χ	pollutant concentration		

zontal temperature distributions affect the wind structure, which in turn regulates the pollution concentration. The most important feedback from the local scale to a larger scale, but yet still subsynoptic, is related to cloudiness and precipitation. Of the synoptic factors the wind exercises the greatest control over the local-scale factors.

The distinction between the urban and surrounding rural environments lies in their entirely different surface structure. In qualitative terms the natural or agricultural landscape is characterized by vegetation and a loose, generally spongy and permeable soil. The urban area, in contrast has a highly compacted, impermeable surface. Obviously these contrasting surfaces have different heat capacities and heat conductivities. This difference is well shown in Table 2.1 by comparison of the physical constants for three surface materials typical of the rural and urban environments. This comparison does not yet reflect the even greater differences that occur when, instead of soils, one looks at humus and plant stands. This is perhaps best illustrated by a comparison of the depths to which the diurnal and annual temperature waves penetrate. Under a light vegetation cover the diurnal temperature wave on a sunny day is completely damped out in a 20–40-cm depth. By contrast, under compacted pavements this wave penetrates to 80–100 cm. The corresponding values for the annual wave are about 4–8 m under a grass cover and 15–19 m under pavement. (A snow cover would greatly alter these values.) In areas with freezing temperatures the recognition of this difference is of great importance for the depth at which water and sewer lines are laid because frost penetrates under traffic arteries, streets, and sidewalks to a much greater depth than indicated by the usual soil temperature measurements, which are made under sod.

TABLE 2.1

Comparison of Physical Constants for Surface Materials Typical of Urban–Rural Environments

Surface	Heat conductivity (cal cm^{-1} sec^{-1} deg^{-1})	Heat capacity[a] (cal cm^{-3} deg^{-1})
Dry soil	6×10^{-5}	8×10^{-1}
Wet soil	5×10^{-3}	5×10^{-1}
Concrete	11×10^{-3}	9×10^{-2}

[a] In lieu of the heat capacity the specific heat is often used (cal gm^{-1} deg^{-1}).

2.2 OBSERVATIONAL PROCEDURES

There are numerous reasons why one wants to ascertain the urban climate, aside from scientific curiosity as to how this climate differs from that of the rural area. Each application requires its own methodology. Many engineering pursuits need climatic data for design purposes. These include building construction, highway and bridge design, drainage systems, stack designs for minimizing effluent nuisances, space heating, and air conditioning and similar pursuits.

It cannot be overemphasized that data from the ordinary Weather Service observing installations at airports are generally not representative for the urban area. In many places there are older records taken in the city that may be used with caution. It will at times be possible to reduce a long airport record to the urban condition by a short comparison series through statistical manipulation. It depends greatly on the particular element. For temperatures and dew points such a procedure may work out satisfactorily, but for wind speed, visibility, and precipitation it is entirely unsatisfactory. Yet a rural reference station close to the urban area for macroclimatic observations is necessary. It should be equipped with recording instruments. A minimum observing program would include: anemometer and wind vane at the standard height of 10 m, dry and wet bulb (or dew point) temperatures, visual range, solar radiation intensity on a horizontal surface, and precipitation recorder. Air-quality control equipment is also highly desirable with the capability of determining particulate concentration, oxidant (O_3), oxides of sulfur and of nitrogen, ammonia, carbon monoxide, and acidity of precipitation.

An analogously equipped station should be set up in the inner city. Where an urban area has several cores multiple meteorological stations are needed. In particular, there is a need for a fairly dense network of precipitation gauges. In the METROMEX project, which studied the rainfall in and around St. Louis, Missouri, one rain gauge per 24 km² (9 miles²) was used in a circle of 42-km (26-mile) radius.

One of the points that must be emphasized is that in any sizeable area, as occupied by a metropolis, there would be microclimatic differences at various sites. These interact with the new microclimates created by human occupation. This differentiation into microclimates is considerably complicated in a complex topographic setting.

Elevation differences are always introducing temperature changes, especially when nocturnal inversions are present. This alone may necessitate a number of smaller substations with thermographs, or at least maximum and minimum thermometers.

In all urban areas the wind conditions are disturbed not only by the buildings but also by altered thermal conditions. This necessitates the need for anemographic equipment. The exposure for such wind measurements is always a difficult problem because of the differences at roof height and at street level. Generally, compromise solutions have to be found. Anemometers above roof level, especially in neighborhoods with uniform building heights, have proved useful. Instrumentation on microwave and broadcast towers at various heights above the ground has been used in a number of investigations, but even then there is some interference with wind-speed values if the wind first blows through the structure before hitting the anemometer. Towers both at the outskirts and in the inner city, if available, should be used for instrumentation, which needs to be installed at least at three levels. As a historical footnote it is interesting that the Eiffel tower in the middle of Paris had meteorological instrumentation installed on it by its constructor, Alexandre Gustave Eiffel (1832–1923) since its completion in 1889 and several publications about the observations have resulted.

Mobile surveys have been an indispensible tool of urban climatology for half a century. Every conceivable vehicle has been instrumented and used: bicycle, automobile, streetcar, and for air pollution observations, elaborate vans. Thermometric observations are usually the minimal objective but equipment to measure humidity and wind speed is usually included. In recent years radiation flux measurements and observations of the infrared radiation temperature of surfaces have become routine in urban climate surveys. In all instances proper care has to be taken to avoid interference of the vehicle, particularly from heat emissions of vehicle engines and exhausts.

Airborne surveys have yielded important data in the third dimension for the air layers above cities. Helicopters, fixed-wing aircraft, radiosonde balloons, captured balloons (kitoons), and tetroons have been used with the whole arsenal of instrumentation, including air sampling. Satellite observations, especially of infrared emission temperatures, have joined the more classical methods. Radar is used in many surveys for the monitoring of precipitation formation in the urban area. For assessments of inversions and particulates in the vertical, laser soundings (lidar) have come into use (Uthe, 1972).

References

Aichele, H. (1968). Über die Verwendung fahrbarer Temperaturschreiber bei geländeklimatischen Untersuchungen. *Z. Angew. Meteorol.* **5**, 267–276.

Landsberg, H. E. (1970). Meteorological observations in urban areas. *Meteorol. Monogr.* **11** (33), 91–99.

Monteith, J. L. (ed.). (1972). "Instruments for Micrometeorology," IBP Handbook No. 22, 263 pp. Blackwell, Oxford.

Uthe, E. E. (1972). Lidar observations of the urban aerosol structure. *Bull. Am. Meteorol. Soc.* **53**, 358–360.

3

Urban Air Composition

3.1 THE URBAN ATMOSPHERE

It has already been stressed that a principal atmospheric difference between urban and rural areas is the drastically altered load of pollutants. To be sure, country air is not entirely a gaseous mixture of N, O, CO_2, Ar, Ne, Kr, Xe, Rn, but there are some traces of O_3 and some solids, principally from soil blowing. Decay products from plant material such as NH_3 and H_2S are also present. However, the urban atmosphere, in addition to the normally present constituents of air, is loaded with admixtures of anthropogenic origin.

Conspicuous among the principal urban atmospheric constituents are the gases SO_2, NO, NO_2, CO, and a number of organic compounds, as well as photochemically produced products including ozone. There is also an overwhelming variety of solids. These are usually designated as aerosols. A partial list includes the elements: Al, As, C, Cd, Cr, Cu, Fe, Mn, Ni, Pb, Ti, V, Zn.

From early times on we find in the literature comments on the dust veil of the cities, and it is not surprising that large cities have been compared to volcanoes. The emission of all these effluents from combustion processes of all kinds does not end the effects. There is a lively chemical interaction taking place. Many of the anthropogenic products are catalysts promoting chemical reactions. Energy is fed into the system by solar radiation and a large number of photochemical reactions take place.

The basic sources are motor vehicles, power production, refineries and various industries, incineration of wastes, and space heating. We can list here only a few of the more common reactions in urban areas and refer the reader to a large volume of specialized literature (Butcher and Charlson, 1972; Heicklen, 1976). Among these reactions is the one leading to ozone (O_3) formation. This results from photochemical splitting of nitrogen dioxide, a product of all high-temperature combustion processes, including those taking place in car engines:

$$NO_2 + h\nu \longrightarrow NO + O \tag{I}$$

$$O_2 + O + M \longrightarrow O_3 + M \tag{II}$$

$$O_3 + NO \longrightarrow NO_2 + O_2 \tag{III}$$

where $h\nu$ represents the solar energy, h being Planck's constant, ν is the frequency of the radiation, and M is a catalyst.

The reaction is represented here as a continuous chain. The ozone, a highly reactive gas, can and will enter into other reactions and become part of the photochemical smog so well known in many major cities. These reactions usually involve a very large number of other components, including such other combustion products as carbon monoxide (CO), sulfur dioxide (SO_2), and hydrocarbons. Atmospheric water vapor or water resulting from combustion of oil, gas, or gasoline also enters into the smog chemistry, often by formation of the OH radical from H_2O.

One important end product involving the nitrogen dioxides and various radicals forms peroxyacetylnitrate (PAN)

$$NO_2 + R\cdot \longrightarrow CH_2COOONO_2$$
$$+ \text{ olefinic compounds or alkyl benzene or acetylene} \tag{IV}$$

PAN is an irritant gas, largely responsible for eye irritation in smoggy atmospheres.

An approximation of a set of chemical reactions involved in the formation of photochemical smog is shown in Table 3.1. The process is actually even more complicated and includes some compounds

TABLE 3.1

Chemical Reactions in Formation of Smog[a,b]

Reactions	Rate constants $(ppm^{-1} min^{-1})$[e]
1. $NO_2 + h\nu \longrightarrow NO + O$	0.355 (min^{-1})
2. $O + O_2 + M \longrightarrow O_3 + M$	2.76×10^6 (min^{-1})[c]
3. $O_3 + NO \longrightarrow NO_2 + O_2$	21.8
4a. $O_3 + NO_2 \longrightarrow NO_3 + O_2$	
4b. $NO_3 + NO_2 \rightleftharpoons N_2O_5$	6×10^{-3}
4c. $N_2O_5 + H_2O \longrightarrow 2HNO_3$	
5. $NO + NO_2 + H_2O \longrightarrow 2HNO_2$	2.5×10^{-3d}
6. $HNO_2 + h\nu \longrightarrow NO + OH\cdot$	5×10^{-3} (min^{-1})
7. $CO + OH\cdot \xrightarrow{O_2} CO_2 + HO_2\cdot$	2×10^2
8. $HO_2\cdot + NO_2 \longrightarrow HNO_2 + O_2$	1×10^1
9. $HC + O \longrightarrow \alpha RO_2\cdot$	3.1×10^4 $(\alpha = 5)$
10. $HC + O_3 \longrightarrow \beta RO_2\cdot + \gamma RCHO$	1.7×10^{-2} $(\beta = 1.9)$
11. $HC + OH \longrightarrow \delta RO_2\cdot + \epsilon RCHO$	1×10^4 $(\delta = 0.2; \epsilon = 0.22)$
12. $RO_2\cdot + NO \longrightarrow NO_2 + \theta OH$	1.8×10^3
13. $RO_2\cdot + NO_2 \longrightarrow$ Products	10
14. $HO_2\cdot + NO \longrightarrow NO_2 + OH\cdot$	1.8×10^3

[a] From Lamb and Seinfeld (1973).
[b] Note: It is possible to combine Reactions 4a–c as

$$O_3 + NO_2 \xrightarrow[\substack{NO_2 \\ H_2O}]{4} 2HNO_3$$

in which the overall rate is that of Reaction 4a, the rate-controlling step.
[c] Pseudo-first order.
[d] Pseudo-second order.
[e] Unless otherwise indicated.

with extremely short lifetimes. This table is taken from Lamb and Seinfeld (1973).

Very important for urban meteorology is also the formation of sulfuric acid from SO_2. With a number of intermediate steps this can be symbolized by

$$SO_2 + O \longrightarrow SO_3 + H_2O \longrightarrow H_2SO_4 \qquad (V)$$

This highly hygroscopic acid is often the cause of low visibility and urban fog formation. It also reacts with ammonia (NH_3) to form a solid aerosol, ammonium–sulfate $(NH_4)_2SO_4$, thus adding to the particulate load of urban air (Tomasi *et al.*, 1975).

The foregoing has barely touched on the complexity of chemical reactions in urban atmospheres. Literally hundreds of transformations go on simultaneously or in sequence, and a whole special science has developed (Tuesday, 1971) to which we can only refer in passing. Aside from the chemically formed particulates, there are other sources of particles. These include soot, fly ash, abrasion fragments, and a variety of chemical fumes from manufacturing processes. The importance of the anthropogenic gaseous and aerosol admixtures to urban air must be viewed in a dual perspective. The first is the actual or potential health effects. These have led to the establishment of ambient air standards. The other perspective is the action on and the interaction with meteorological elements. Only the latter will be discussed in detail in this book.

3.2 AIR STANDARDS AND OBSERVATIONS

In the United States the air quality standards are set by the Clean Air Act and its implementing federal regulations. These cover only the principal pollutants: Total suspended particulates (TSP), SO_2, CO, NO_2, photochemical oxidants (principally O_3), and nonmethane hydrocarbons (HC). The primary standards are essentially mandatory targets. These are to be reached by emission controls. There are also some secondary standards that constitute the ultimate desired limit of pollutants. Table 3.2 lists the presently valid primary United States air quality standards. The hourly values listed in the table are not to be exceeded more than once per year.

As with many other standards, the desired values are occasionally exceeded. This can be caused in many instances by weather conditions, and these are described as stagnation situations. They are characterized by a stationary high-pressure pattern with a ground- or low-level temperature inversion and calm or very light wind. This type of weather is depicted in Fig. 3.1. Under those circumstances the public must be alerted to the health dangers, and in extreme cases sources of the pollutants have to be shut down. Table 3.3 gives a pollution standard index of the U. S. Environmental Protection

TABLE 3.2

Primary United States Ambient Air Quality Standards[a]

Pollutant	Averaging time	Weight (μg m^{-3})	ppm
Particulates	Geometric mean annual	75	—
	24 hr	260	—
Sulfur oxides	Arithmetic mean annual	80	0.03
	24 hr	365	0.14
Carbon monoxide	8 hr	10 (mg m^{-3})	9
	1 hr	40 (mg m^{-3})	35
Nitrogen dioxide	Arithmetic mean annual	100	0.05
Photochemical oxidants	1 hr	160	0.08
Nonmethane hydrocarbons	3 hr (6–9 a.m.)	160	0.24

[a] Set by federal regulations in 1975.

Agency, together with the danger level to health and the corresponding pollution levels.

It should be noted that the various pollutants need not behave the same way in various urban areas, although in the aggravated cases they all ordinarily exceed the standards. This is shown in Table 3.4 for observations in two cities in 1975.

Comparisons among urban areas in this respect are very difficult. One city may exceed the standards for one pollutant and meet them for another, and another city may show just the opposite behavior. For example, in 1974 Chicago had a very good value for TSP (44 μg m^{-3}) and a poor one for NO_2 (133 μg m^{-3}). In contrast, Steubenville had a high value for TSP (116 μg m^{-3}) and a lower one for NO_2 (98 μg m^{-3}). These conditions reflect both emission factors and weather. Even from year to year the values vary greatly. That is generally a reflection of the weather conditions. Hence it is difficult to detect trends in the observed values. A cautious appraisal of the data indicates a slight downward trend in United States cities in TSP, SO_2, and CO in the last decade thanks to the control measures, but the NO_2 values seem to stay steady or even show local increases (Council on Environmental Quality, 1977). A time series for SO_2 and TSP is shown in Fig. 3.2.

Another cautionary statement is necessary here. All urban areas show wide ranges in air quality. Land use, traffic density, and distribution of major point sources, such as power plants, play a role in

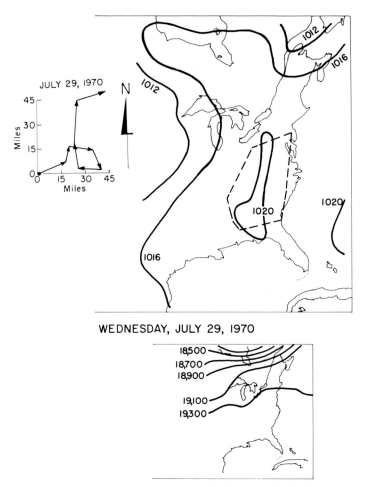

Fig. 3.1a Synoptic situation indicating a stagnation pattern leading to accumulation of air pollutants. Upper panel indicates surface isobars (mbar) with an inset on left showing composite of three hourly wind vectors at Washington, D. C. The dashed line indicates area for which an air pollution alert was issued. The lower panel shows the 500-mbar heights (ft), indicating no gradient over the alert area.

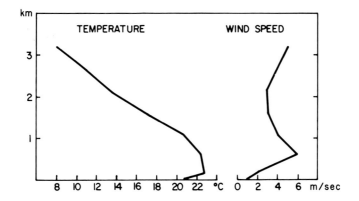

Fig. 3.1b Upper-air sounding corresponding to weather situation on July 29, 1970 shown in Fig. 3.1a at Sterling, Virginia, the sounding station near Washington, D. C.

the emission arrangement. The city structure itself acts as a principal control of the dispersion. One can readily see this on individual days. A particularly interesting case for an April day has been three-dimensionally placed on record for the highly industrialized area of Mannheim–Ludwigshafen by Georgii (1969). Figure 3.3a shows the horizontal distribution of SO_2 above that region. At that height the surface inhomogeneities in concentrations should have

TABLE 3.3

Pollution Standard Index and Health Effects[a]

Index value	Quality level	24 hr TSP	24 hr SO_2	8 hr CO	1 hr O_3	1 hr NO_2	Health designation
100	Standard	260	365	10	160	160	Can be lethal for old and ill persons
200	Alert	375	800	17	400	1130	Hazardous
300	Warning	825	1600	34	800	2260	Very unhealthy
400	Emergency	875	2100	46	1000	3000	Unhealthy
500	Significant harm	1000	2620	575	1200	3750	Intolerable

[a] All values are given in $\mu g\ m^{-3}$ except for CO, which is given in $mg\ m^{-3}$.

TABLE 3.4

Frequency (in Days during 1975) of Exceeding Pollution Standards Index (PSI) in St. Louis (St. L) and Denver (D)

Pollutant	PSI					
	100–200		200–300		>300	
	St. L	D	St. L	D	St. L	D
CO	44	57	6	40	4	0
O_3	74	73	6	0	1	0
TSP	3	5	1	2	1	0

essentially disappeared. In Fig. 3.3b the vertical distribution to 700 m above the surface is shown. The rapid drop-off from the restricted area of emission is notable. Still, urban plumes can often be traced several hundred kilometers downwind (Georgii, 1978).

In a climatological sense a rapid drop-off is also readily seen in the mean values for an urban area. In the metropolitan area of Washington, D. C., the mean values are fairly closely related to the density of built-up sections. Because of the lack of industry the SO_2 pollution must be attributed to space heating, transportation, and several power plants (Fig. 3.4).

Raynor *et al.* (1974) have analyzed the drop-off of SO_2 pollution downwind from New York City over Long Island. Their results are

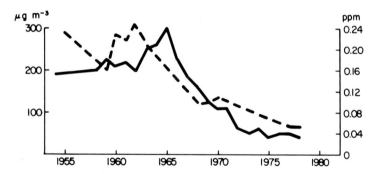

Fig. 3.2 Change of pollutant levels in United States urban areas prior to and after enactment of Clean Air Act. Left scale is for total suspended particulates (TSP) (dashed line); right scale is for SO_2 (solid line).

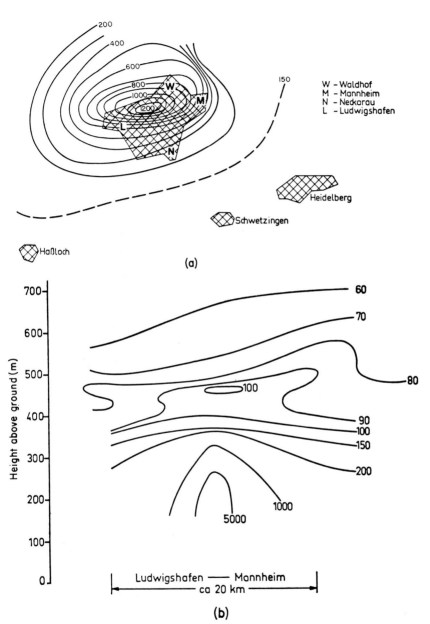

(a)

(b)

Fig. 3.3 Horizontal (a) and vertical (b) distribution of SO_2 concentration (μg m^{-3}) in the Mannheim–Ludwigshafen, West Germany, industrial area (after Georgii, 1969).

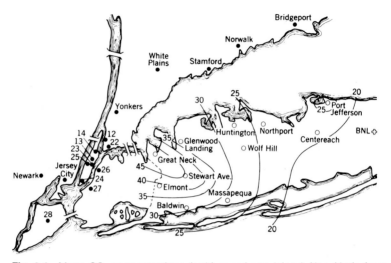

Fig. 3.4 Mean SO$_2$ concentrations (ppb) over Long Island, New York, from 2 yr of observation (1968–1969). Solid circles are New York City monitoring stations. Data were used from stations marked by open circles (from Raynor *et al.*, 1974).

shown in Fig. 3.5, which indicates essentially a logarithmic decrease in concentration at the surface. This is consistent with a number of other sets of observations elsewhere.

Actually, the pollutants singled out for determination of air pollution standards hardly characterize the extent of the urban pollution problem. Nottrodt *et al.* (1980) undertook an analysis of trace elements in Frankfurt, West Germany. This included many of the metallic elements. Some of these, such as iron, act as the catalysts in chemical reactions; others like vanadium and lead can be serious health hazards. But many other elements are abundant in the earth's crust and are common in blowing soil. The study confirmed that most of these elements fluctuate similarly as a function of time, under the influence of changing weather. The correlation coefficients between concentrations of trace metals were between 0.72 and 0.83. Auto exhaust products, such as lead and bromine, decreased by a factor of 10 from street to roof level. All authropogenic pollutants decreased by an order of magnitude between Frankfurt and the 800-m-high Tannus Mountains, 25 km distant, and by two orders of magnitude to the clean air of the Seviso mountains. The natural elements, such as calcium, stayed about the same.

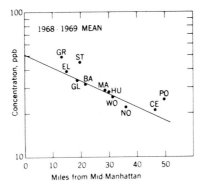

Fig. 3.5 Logarithmic drop-off of SO$_2$ concentration (ppb) (1968–1969 mean) over Long Island with distance from New York City (from Raynor *et al.*, 1974): GR, Great Neck; EL, Elmont; GL, Glenwood Landing; ST, Stewart Avenue; BA, Baldwin; MA, Massapequa; HU, Huntington; WO, Wolf Hill; NO, Northport; CE, Centereach; PO, Port Jefferson.

An analysis of street dust in Urbana, Illinois (Hopke *et al.*, 1980) for about a third of the elements in the periodic table showed the concentrations given in Table 3.5. The sample, which at present is only indicative of the trace elements, was gathered at a street intersection with moderate traffic density of about 7000 cars/day. As sources one can enumerate: soil, cement, tire wear, vehicle exhaust, and salt residues from snow-removal efforts. The particles ranged in size from 20 to 500 μm and one must suspect that in strong winds a substantial fraction would be suspended in the air. From an hygienic point of view the composition of this dust is not encouraging.

Table 3.5 refers principally to anorganic elements. However, the organic compounds are of greatest interest for public health. Among them are the polycyclic hydrocarbons, which are known as carcinogenic entities. Their concentration in urban areas seem to be related to motor vehicle traffic, and their fluctuations are parallel to those of particulates (Handa *et al.*, 1980).

The anthropogenic component in urban pollution is clearly shown by its weekly cycle. Lincoln and Rubin (1980) have clearly documented that in Allegheny County, Pennsylvania, which includes the city of Pittsburgh. Both TSP and CO are about 30 percent less downtown on Sundays than on weekdays. The correlation of these elements to day of the week for a year of observations was 0.98. The lowest values were for Sunday, followed by Saturday. The highest

TABLE 3.5

Concentration of Elements in Urbana, Illinois, Street Dust[a]

Element	μg/g	Element	μg/g
Manganese	350 ± 30	Uranium	3.5 ± 0.7
Zinc	320 ± 30	Samarium	3.4 ± 0.5
Barium	310 ± 54	Calcium	2.7 ± 0.5%
Nickel	250 ± 60	Antimony	2.2 ± 0.3
Strontium	250 ± 50	Cadmium	1.6 ± 0.2
Chromium	210 ± 20	Dysprosium	1.6 ± 0.2
Zirconium	120 ± 14	Cesium	1.1 ± 0.2
Bromine	84 ± ?	Selenium	1.0 ± 0.3
Cerium	29 ± 1	Ytterbium	1.0 ± 0.2
Rubidium	29 ± 5	Potassium	0.94 ± 0.13%
Arsenic	11 ± 1	Sodium	0.53 ± 0.05%
Lanthanum	10 ± 1	Terbium	0.44 ± 0.13
Cobalt	6.8 ± 0.4	Europium	0.4 ± 0.03
Iron	6.2 ± 0.5%	Silver	0.2 ± 0.09
Hafnium	5.0 ± 0.5	Lutetium	0.16 ± 0.4
Gallium	4.9 ± 0.9	Lead	0.1 ± 0.02%
Thorium	4.3 ± 0.3	Mercury	0.09 ± 0.008
Scandium	4.2 ± 0.3		

[a] After Hopke *et al.* (1980).

values were observed on Monday, followed by Friday, marking the influx to and exodus from the city. In this area the rural locations showed only 4 percent of the urban concentration of the pollution caused by motor vehicles.

3.3 POLLUTANT–WEATHER INTERACTIONS

Weather conditions affect pollutant concentrations and there are some effects of pollutants on weather. Simplest in this respect are the diurnal and annual variations. These are, of course, in part governed by changes in emissions. In the diurnal cycle, in typical American cities, there is a notable double peak of emissions caused by the traffic during the morning and evening rush hours. The morning maximum is more pronounced because traffic is usually more

condensed in time, and vertical mixing at that time is small. Facto-
ries, businesses, and homes will also start their daily work and
heating cycles at that time. All this coincides with a climatological
minimum of surface wind speed. During midday the surface inversion is usually eliminated and in warmer seasons and climates replaced by convection. This raises the mixing height, which is defined as the thickness of the layer of air from the ground up to the
level at which vigorous vertical mixing occurs. The temperature decrease with height in this layer is adiabatic (1°C/100 m) or even su-
peradiabatic.

Some typical examples of the low-level stability conditions in
urban areas are shown in Figs. 3.6 and 3.7. The former gives the percent frequency of ground inversions near Denver, Colorado, as a
function of time of day, by seasons. The high frequencies in the
early morning hours and at night are remarkable. In summer and
spring these rarely persist through midday, but in winter this condition will often persist throughout the day (Moore, 1975). This is not
unique, but is found in most urban areas in higher latitudes. The per-
sistence of low-level inversions in the cold season is most pronounced in valleys or topographic troughs. Figure 3.7 is taken from
a paper by Popovics and Szepesi (1970) and shows the monthly
average values of the mixing height in Budapest, Hungary, for the
early morning and afternoon. It also indicates the difference
between town center, suburban, and rural conditions. The graphs
not only show notable seasonal variations but also indicate that the
midday differences in the three settings are negligible. They also
convey the fact that a ground inversion always exists in the rural
area in the morning but that the inversion height is lifted off in the
town center. This is more evident in the cold months than in the
warm season. Heating in the urban area, as we will see in detail
later, contributes to the lifting of the nocturnal mixing height.

The consequence for the pollutants is quite obvious. Midday values are invariably lower for most of them than are early morning
concentrations, especially in summer. The notable exception to the
rule are the oxidants, mainly O_3, which, as we have seen, is photochemically formed. Sunshine is, of course, a prerequisite. An example of the diurnal variation of oxidants in the Washington, D. C.,
center is shown in Fig. 3.8. Depicted is the buildup of the photochemically produced pollutants in a section with heavy automobile

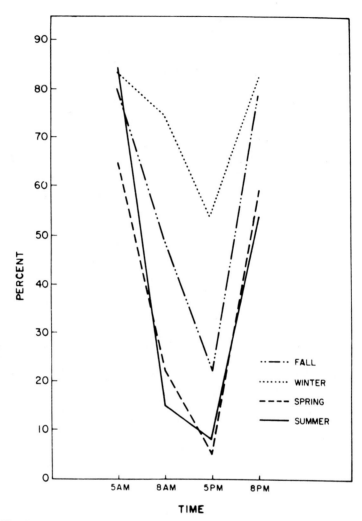

Fig. 3.6 Percent frequency of low-level inversions as related to time of day and season at Denver, Colorado (after Moore, 1975).

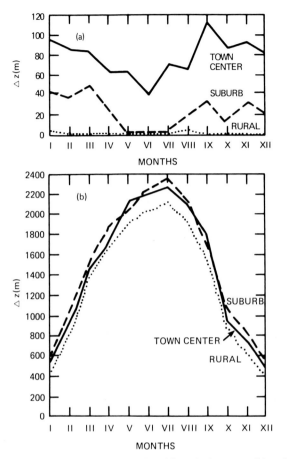

Fig. 3.7 Mixing heights in the morning (a) and afternoon (b) at Budapest, Hungary, in their annual course (after Popovics and Szepesi, 1970).

traffic during the midday hour and its rapid decay after sunset. The observations were taken during the same stagnation situation shown earlier in Fig. 3.1.

The most obvious effect of pollutants is on the visual range. All population centers have less visibility than nearby rural areas. In the United States the difference shows, to give a rough estimate, 10–20 percent more observations with low visibilities (<10 km) in urban centers than in the countryside. In some European industrial centers the difference is even higher. The greatest increase is in the inci-

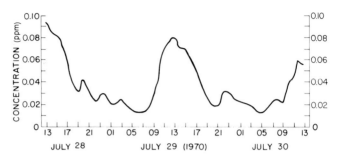

Fig. 3.8 Oxidant concentrations at Union Station, Washington, D. C., during stagnation episode, July 28–30, 1970.

dence of fog with visibilities <1 km, which may even be twice as high in the urban areas compared with rural environs.

The relation is in fact quite complex. Although the daytime visibility V is inversely proportional to the extinction coefficient σ of the aerosol

$$V = \frac{3.912}{\sigma} \ (\text{km}) \tag{3.1}$$

assuming a normal visual contrast $\epsilon = 0.02$, is not simply expressible by easily obtained characteristics of the aerosol. The extinction, a wavelength-dependent factor, can be readily measured if it varies from 0.02^{-1} km in very clean air to around 0.5^{-1} km in polluted air. For a dry dust aerosol the extinction of light, which governs the visibility, in spherical particles is

$$\sigma \sim \frac{N}{\pi r^2} \tag{3.2}$$

where N is the number of particles per cubic centimeter and r the radius of the particles, usually given in micrometers.

The complications are immediately obvious. Solid particles in the atmosphere are rarely spherical. There is a wide spectrum of radii present at any one time and the coefficient is different for opaque material and droplets. Hence it is a futile attempt to develop theoretical models relating urban visibilities to physical parameters of the aerosol.

Empirically many facts are known. The diurnal variation is an inverse function of the emission strength. Hence, the early morning hours have the lowest visual ranges, except for those localities where photochemical smog is frequent. Similarly, winter is usually the season of lowest visibility in the low atmospheric layers but with high photochemical visibility reductions in the early afternoons in summer.

For some of the pollutants, such as SO_2 turning into sulfuric acid, visual ranges depend greatly on the ambient relative humidity because of the hygroscopicity of this pollutant. It begins to grow into droplets interfering with visibility at 60 percent relative humidity, and the droplets grow rapidly when humidities are higher. Observed visual ranges as related to SO_2 concentrations at high humidities are shown in Fig. 3.9. There is little doubt that this aerosol is one of the major culprits in the formation of urban fogs.

The statistics clearly show increases in hours of low visibilities from the 1930s to the 1960s in many urban areas of the United States and industrialized central and western Europe. With the inaugura-

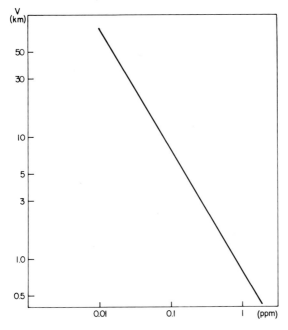

Fig. 3.9 Relation of visibility to SO_2 concentration at high relative humidities ($\geq 90\%$).

tion of clean air acts, there has been a measurable decrease in both suspended material and SO_2 with resultant improvements in visibility. A typical example of the vast improvements achievable are shown in Fig. 3.10 for Pittsburgh in a 25-yr interval (Davidson, 1979).

Closely related to effects of pollutants on visibility are their influence on the attenuation of incoming solar radiation. This is most pronounced on the short wavelengths, especially the ultraviolet. On many days in polluted areas this part of the solar spectrum is almost eliminated. This is particularly true for low solar elevations, but even the other wavelengths are severely attenuated. Much of this scattered radiation gets back to earth as sky light. It causes the pale sun and the murky sky so characteristic of polluted urban atmospheres.

A number of investigations have dealt with rural–urban differences in radiation, many of them referring to typical days. Even on relatively "clean" days this difference is very unstable. Figure 3.11 shows such a case for the St. Louis area (Bergstrom and Pe-

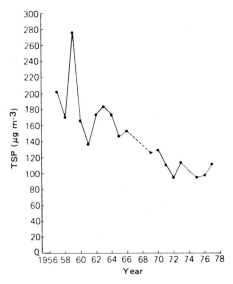

Fig. 3.10 Average values of TSP ($\mu g\ m^{-3}$) at County Office Building in downtown Pittsburgh, Pennsylvania (from Davidson, 1979).

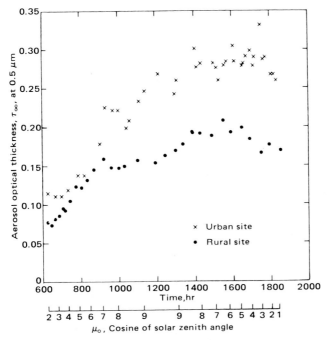

Fig. 3.11 Daily variation of measured optical thickness at 0.5 μm on August 10, 1972, at an urban and a rural site in the St. Louis area (from Bergstrom and Peterson, 1977).

terson, 1977) in terms of the optical thickness for aerosols, defined as

$$\tau = \int z(k + \sigma)\,dz \qquad (3.3)$$

where

k absorption coefficient
σ scattering coefficient
z height of the scattering layer

The midday values of the urban site are twice those in the rural area.

Similar comparisons have been made for Milwaukee, where on some individual days the extinction coefficients were measured for a wavelength of 500 nm in the clean air over Lake Michigan and at urban sites (Bridgman, 1979). Over the lake, σ values of 0.018 km^{-1} prevailed, whereas over the rail yards a maximum of 0.052 km^{-1} was observed. On the day of these measurements (April 12, 1976) the lake, urban, and rural values, above an inversion, became the same. Such values are probably quite typical and clearly illustrate the differences between clean and polluted air.

For global radiation, i.e., the total radiation incident at the surface from sun and sky, there are a number of climatological comparisons. Terpitz (1965) compared the measurements in the metropolis of Cologne (1 million inhabitants) and the town of Trier (75,000 inhabitants) in the same climatic area. Cologne is heavily industrialized with much vehicular traffic. The area around the Trier weather station consists of vineyards and the town has little industry. In midwinter the Cologne global radiation is 10 percent less than in Trier. In midsummer, the reduction is only about 3 percent.

Secular variations in turbidity could be followed in Jerusalem, where population growth, industrialization, and motor vehicle traffic have increased somewhat in recent years but not excessively (Joseph and Manes, 1971). Spectral measurements of turbidity between 1930–1934 and 1961–1968 showed increases of 30 percent or about 10 percent per decade. Similar increases occurred in Washington, D. C., but Mexico City had decadal increases of 25 percent per decade. In Jerusalem regional effects may have been more significant than changes in the city itself. However, as in Washington and Mexico City, photochemically formed aerosols from motor vehicle exhausts can readily account for the observed increases in turbidity.

In the foregoing the effect of pollutants on meteorological elements was the focus of attention. The opposite is also important, the effect of meteorological elements on pollutant concentration. Two of them are active: precipitation and wind. The precipitation effect is well marked, especially on the particulates. This is also a twofold effect, designated as rainout and washout. In the former, the aerosols act as nuclei of condensation and are carried with the raindrops to the ground. In washout the aerosols are carried down after collision with the raindrops or snowflakes. For urban air, washout is the more important cleansing process. In hefty showers aerosol concentrations may be cut in half in 15 min, a process that may take 2 hr in a slow, steady rain. There are again many variables at work: total rain amount, spectral distribution of droplet size, rain duration, and the continuing emission strength. As a first approximation one can represent the process by

$$\frac{d\chi_t}{dt} = \chi_0 I_p e^{-wt} \qquad (3.4)$$

where

χ_0, χ_t aerosol concentrations at times t_0 and t, respectively
I_p intensity of precipitation (mm/hr)
w washout coefficient (which is dependent on drop-size spectrum)

Observations on Aitken nuclei (diameters about $0.01-0.1$ μm) showed reductions of 12 percent after rain in urban atmospheres (Landsberg, 1938). In air with initially 77.000 Aitken nuclei cm^{-3}, Georgii (1965) found a reduction of only 10 percent for a slow rain of 2 mm hr^{-1}. Laboratory experiments (Beilke, 1970) simulated this event with the same intensity and mean raindrop diameters of 0.62 mm. After 70 min the initial nuclei concentration of 77,000 cm^{-3} fell to 69,000 cm^{-3}. Similar simulation of SO$_2$ concentrations at 0.5 mm hr^{-1} intensity and mean raindrop diameters of 0.46 mm showed in 42 min a fall to half the original concentration.

Wind is the best diluter of urban contaminants. Figure 3.12 is drawn from a number of diverse empirical observations: The SO$_2$ concentrations as a function of wind speed were measured in Halle, East Germany (Noack, 1963); the dust weights were measured in the Russian city of Charkov (Sheleikovskii, 1961); the CO was measured at a 3-m height in a busy street in Frankfurt, West Germany (Georgii, 1969); and the TSP were climatological mean values measured in several large United States cities in 1974. All elements show sharp drop-offs with wind speed. Obviously there is no uniform law governing the conditions, which are also importantly influenced by other variables. In general, however, one can state that

$$\chi_{(u)} = c(Q_0/\bar{u}) \qquad (3.5)$$

where

χ_u concentration of the pollutants (μg m^{-3})
Q_0 area emission strength (μg m^{-2} sec^{-1})
$\bar{u}$ mean wind speed for a given time interval (m sec^{-1})
c complex dimensionless parameter of stability

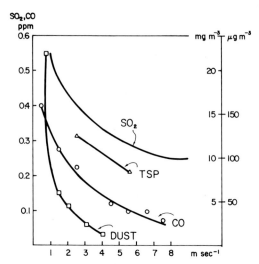

Fig. 3.12   Relation of pollutant concentration to wind speed: left scale (ppm) refers to SO₂ and CO; right scale refers to dust (mg m⁻³) and TSP (μg m⁻³).

It needs to be emphasized here that urban wind speeds vary in a complicated manner horizontally and vertically, depending on building heights and street layout, as we shall see in more detail later.

3.4 URBAN CLIMATOLOGICAL DISPERSION MODELS

There are a number of quasi-quantitative representations of the urban boundary layer for the purpose of assessing the dispersion of pollutants. Most of them are designed for diagnostic and predictive purposes for short time scales. They serve the daily needs of air pollution control and warning systems. The nature of the problem is such that none of them are very precise. The correlations between predicted and observed pollution concentrations indicate that the models usually explain less than 60 percent of the variance (Gifford, 1974). The complexities are already pronounced for inert substances, but they become quite overwhelming when the chemical reactions have to be incorporated (Lamb and Seinfeld, 1973). These models not only reflect the difficulty of representing the atmospheric turbulence but also the ambiguities of the large number of reactions and their partially uncertain reaction rate constants. Although these mathematical representations have some merit for instantaneous weather conditions, they have little bearing on the pollution climate of an urban area.

The development of climatological models has aimed at estimates of pollutant concentrations for longer intervals of time, such as monthly, seasonal, or annual values. These are principally applicable to chemically relatively inactive species such as carbon monoxide (CO) and particulates. Other species have been incorporated by assigning an empirically determined half-life for a specific pollutant. All of these models are based on the basic model of turbulent diffusion. This is usually referred to as the *Gaussian model,* because the lateral and vertical dispersion at a given distance from the source represent a random frequency distribution. This is determined as time-averaged concentrations around an axis specified by the mean wind direction (Gifford, 1974).

For urban models this is established from the emissions in a given unit area moving to a receptor area. In a climatological analysis the wind frequencies are presented for a 16-point wind rose, making each sector 22.5°. The average concentration χ_A for an area source is then given for a particular receptor by

$$\bar{\chi}_A = \frac{16}{2\pi} \int_0^\infty \left[\sum_{k=1}^{16} q_k(r) \sum_{v=1}^{6} \sum_{s=1}^{6} f(k, v, s) S(r, z, \bar{u}, P) \right] dr \quad (3.6)$$

where

k identifies the wind sector, 1–16

r distance of source to receptor

v wind-speed class, 1–6

s stability class, 1–6

q_k emission for kth sector; $q_k = \int Q(r, t)$ with Q as emission rate per unit area and unit time t

$S(r, z, \bar{u}, P)$ dispersion function, where z is the height of the receptor above ground.

TABLE 3.6

Pasquill Stability Classes

Surface wind speed (10-m ht) (m sec^{-1})	Day, insolation			Night	
	Strong	Moderate	Slight	Thinly overcast or ≥4/8 low cloud	≤3/8 cloud
<2	A	A–B	B		
2–3	A–B	B	C	E	F
3–5	B	B–C	C	D	E
5–6	C	C–D	D	D	D
>6	C	D	D	D	D

For a ground-level receptor $z = 0$, a common assumption for practical purposes in an urban area is

$$S(r, 0, \bar{u}, P)$$

$$= \frac{2}{(2\pi)^{1/2}\, \bar{u}\sigma_z(r)} \exp\left[-\frac{1}{2}\left(\frac{H}{\sigma_z(r)}\right)^2 \right] \exp\left(\frac{-0.692r}{\bar{u}T_{1/2}}\right) \quad (3.7)$$

where $\sigma_z(r)$ is the vertical dispersion; $\sigma_z(r) = ar^b$, with a and b parameters depending on atmospheric vertical stability. The stability is usually estimated in six classes, originally proposed by Pasquill (1961). These are based on a few commonly observed meteorological elements, namely wind speed and insolation conditions, as shown in Table 3.6. In this scheme D designates neutral stability

TABLE 3.7

Wind Speeds Related to Wind-Speed Classes

Speed class	Class center (m sec^{-1})
1	1.50
2	2.46
3	4.47
4	6.93
5	9.61
6	12.52

characterized by an adiabatic vertical lapse rate (1°C/100 m). The classes E and F are stable inversion conditions and A, B, and C have various rates of instability.

A modified development by Gifford (1961), later adapted for computer use (Busse and Zimmerman, 1973), gave stability classes that could be related not only to wind speed but also to mixing heights. This element gives an idea as to what level in the atmosphere surface pollutants are readily transported by atmospheric motions. In the concentration formula (3.6) the climatological frequencies of wind speed and stability classes appear as summations. These classes and their characteristics are given in Tables 3.7 and 3.8.

The mixing heights and wind speeds give a fairly good estimate of the climatological potential at various localities. The mixing heights are derived from the radiosonde ascents made twice daily at many locations. Wind speeds refer to the conventional 10-m height. It must be noted that these observations are made at airports and that wind speeds are generally lower in the built-up areas. Average conditions for a few selected cities in the United States are shown in Table 3.9. The diurnal variation of both mixing heights and wind speeds are clearly shown. Both elements show higher values in the afternoon. The high air-pollution potential in the Great Basin and the pacific Southwest is clearly shown. In general, towns in valley or trough locations are most afflicted by poor dispersion conditions.

A somewhat different climatological approach has been used in an assessment of the quantity C in Eq. (3.5) given in Section 3.3. That dimensionless parameter is essentially a function of stability in that it depicts the ratio of the width DX of the urban area to the depth of the pollutant cloud. By use of the vertical dispersion constants, indicated in Eq. (3.7), the parameter C can be defined as

$$C = (2/\pi)^{1/2}\{(DX/2)^{1-b}/[a(1 - b)]\} \qquad (3.8)$$

Hanna (1978) utilized data for a number of stations in Maryland, New Jersey, and Colorado to estimate the diurnal variability of the stability parameter from the ratio $\chi\bar{u}/Q$. The value of C is nearly constant from 6 a.m. to 6 p.m., both in summer and in winter. It is about five times higher in the predawn hours than in midafternoon. For typical values of the parameters a and b in the dispersion factor (Smith, 1968), C will be: 600 for stable conditions, 200 for neutral stability, and 50 for unstable atmospheric stratification.

TABLE 3.8

Stability Classes, Mixing Heights, and Dispersion Functions[a]

Stability	Class (Pasquill)	Mixing Ht/σ_z parameters	Distance (m)							
			100–500		500–5000		5000–50,000			
			a	b	a	b	a	b		
1	A	1.5 × HT	0.0383	1.2812	0.0002	2.0886	—	0.9109		
2	B	HT	0.1393	0.9467	0.0494	1.1137	—	—		
3	C	HT	0.1120	0.9100	0.1014	0.9260	0.1154	0.5642		
4 (Day)	D	HT	0.0856	0.8650	0.2591	0.6869	0.7368			
4 (Night)		(HT + HMin)/2								
5	E	HMin	0.0818	0.8155	0.2527	0.6341	1.2969	0.4421		
6	F	HMin	0.0545	0.8124	0.2017	0.6020	1.5763	0.3606		

[a] The values for the mixing height (HT) in this table can be obtained from tabulations by Holzworth (1972); HMin stands for minimum mixing height.

TABLE 3.9

Average Mixing Heights and Wind Speeds (Winter and Summer) on Days without Precipitation and Seasonal Frequency of Episodes of High Air-Pollution Potential[a] Lasting Two or More Days in a Five-Year Period at Selected Urban Locations[b]

Location	Time	Winter H (m)	Winter $\bar{U}$ (m sec^{-1})	Summer H (m)	Summer $\bar{U}$ (m sec^{-1})	Number of episodes	Season[c]
Albuquerque, NM	a.m.	345	3.7	560	3.6	20	W
	p.m.	1402	5.5	3902	5.8		
Denver, CO	a.m.	178	4.5	243	3.6	10	W
	p.m.	1357	5.5	3358	5.9		
Miami, FL	a.m.	654	5.4	1041	4.3	0	—
	p.m.	1208	6.4	1360	5.3		
New York, NY	a.m.	875	8.3	662	5.5	3	A
	p.m.	901	8.2	1512	6.8		
Oklahoma City, OK	a.m.	296	6.9	344	6.9	0	0
	p.m.	804	8.1	1830	7.2		
Salt Lake City, UT	a.m.	254	3.7	198	4.4	35	W
	p.m.	808	4.0	3673	6.0		
San Diego, CA	a.m.	468	2.0	531	2.0	139	S
	p.m.	989	4.2	564	4.1		
Santa Monica, CA	a.m.	176	2.8	552	1.9	63	W
	p.m.	863	4.4	601	5.1		
Tucson, AZ	a.m.	216	4.3	335	3.8	4	W
	p.m.	1390	5.0	3040	5.3		
Washington, DC	a.m.	539	5.3	378	3.1	9	W
	p.m.	963	6.7	1884	5.4		

[a] Mixing height, $\leq$1000 m; wind speeds, $\leq$1 m sec^{-1}.

[b] Adapted from Holzworth (1972). Useful information on low-level vertical stability and wind speeds for the United States is contained in an atlas of climatological rawinsonde summaries by Holzworth and Fisher (1979).

[c] W, winter; A, autumn; S, summer.

References

Beilke, S. (1970). Untersuchungen über das Auswaschen atmosphärischer Spuren-stoffe durch Niederschläge. *Ber. Inst. Meteorol. Geophys.*, Univ. of Frankfurt, No. 19, 61 pp.

Bergstrom, R. W., and Peterson, J. T. (1977). Comparison of predicted and observed solar radiation in an urban area. *J. Appl. Meteorol.* **16,** 1107–1115.

Bridgman, H. A. (1979). Aerosol extinction at 500 nm in urban and rural air at Milwaukee in April 1976. *J. Appl. Meteorol.* **18,** 105–116.

Busse, A. D., and Zimmerman, J. R. (1973). User's guide for the climatological dispersion model. *U. S. Environ. Prot. Agency, Environ. Monitoring Ser. EPA-R4-73-024,* 131 pp. Research Triangle Park, NC.

Butcher, S. S., and Charlson, R. J. (1972). "An Introduction to Air Chemistry," 241 pp. Academic Press, New York.

Davidson, C. I. (1979). Air pollution in Pittsburgh: A historical perspective. *J. Air Pollut. Control Assoc.* **29,** 1035–1041.

Georgii, H.-W. (1965). Untersuchungen über Ausregnen und Auswaschen atmosphärischer Spurenstoffe durch Wolken und Niederschlag. *Ber. Dtsch. Wetterdienstes* **14** (100), 23 pp.

Georgii, H.-W. (1969). The effects of air pollution on urban climates. *Bull. WHO* **40,** 624–635.

Georgii, H.-W. (1978). Die lokale und regionale Immissionsbelastung der Troposphäre. *Ber. Bunsenges. Phys. Chem.* **82,** 1185–1188.

Gifford, F. A. (1961). Uses of routine meteorological observations for estimating atmospheric dispersion. *Nucl. Saf.* **2,** 47–51.

Gifford, F. A., Jr. (1974). Further comparisons of urban air pollution models. *Rept. to 5th Mtg. NATO/CCMS Panel on Air Pollution Modeling, Roskilde, June 4–6, 1974,* 6 pp. Environ. Res. Lab., NOAA, Atomic Turbulence and Differsion Lab., Oak Ridge, Tennessee.

Handa, T., Kato, Y., Yamamura, T., and Ishii, T. (1980). Correlation between the concentrations of polynuclear aromatic hydrocarbons and those of particulates in an urban atmosphere. *Environ. Sci. Technol.* **14,** 416–422.

Hanna, S. R. (1978). Urban modelling of inert substance. *In* "Air Quality Meteorology and Atmospheric Ozone" (A. L. Morris and R. Barras, eds.), pp. 262–275. Am. Soc. for Testing Materials, Philadelphia.

Heicklen, J. (1976). "Atmospheric Chemistry," 406 pp. Academic Press, New York.

Holzworth, G. C. (1972). Mixing heights, wind speeds, and potential for urban pollution throughout the contiguous United States. *U. S. Environ. Prot. Agency, Publ. AP-101,* 146 pp. Research Triangle Park, NC.

Holzworth, G. C., and Fisher, R. W. (1979). Climatological summaries of the lower few kilometers of rawninsonde observations. *U. S. Environ. Prot. Agency, EPA-600/4-79-026,* 141 pp. Research Triangle Park, NC.

Hopke, P. K., Lamb, R. E., and Matusch, D. F. S. (1980). Multielemental characterization of urban roadway dust. *Environ. Sci. Technol.* **14,** 164–172.

Joseph, J. H., and Manes, A. (1971). Secular and seasonal variations of atmospheric turbidity in Jerusalem. *J. Appl. Meteorol.* **10**, 453–462.

Lamb, R. G., and Seinfeld, J. H. (1973). Mathematical modeling of urban air pollution (general theory). *Environ. Sci. Technol.* **7**, 253–261.

Landsberg, H. E. (1938). Atmospheric condensation nuclei. *Ergeb. Kosm. Phys.* **3**, 155–252.

Lincoln, D. R., and Rubin, E. S. (1980). Contributions of mobile sources to ambient particulate concentration in a downtown urban area. *J. Air Pollut. Control Assoc.* **30**, 777–781.

Moore, D. A. (1975). Clean air prospects for Colorado. *In* "On Some Environmental Problems in Colorado" (O. M. Essenwanger, ed.) (unpaginated). Dept. of Atmospheric Sci., Colorado State Univ., Fort Collins, CO.

Noack, R. (1963). Untersuchungen über Zusammenhänge zwischen Luftverunreinigung und meteorologischen Faktoren. *Z. Angew. Meteorol.* **4**, 299–303.

Nottrodt, K. H., Georgii, H. W., and Groeneveld, K. O. (1980). Temporal and spatial differences in the composition of atmospheric aerosols. *Sci. Total Environ.* **14**, 113–128.

Pasquill, F. (1961). The estimation of the dispersion of windborne material. *Meteorol. Mag.* **90**, 33–49.

Popovics, M., and Szepesi, D. J. (1970). Diffusion climatological investigations in Hungary. *Proc. Int. Clean Air Congr., 2nd 1970,* Paper No. ME-20B, 24 pp.

Raynor, G. S., Smith, M. E., and Singer, I. A. (1974). Temporal and spatial variation in sulfur dioxide concentrations on surburban Long Island, New York. *J. Air Pollut. Control Assoc.* **24**, 586–590.

Sheleikovskii, G. V. (1961). Smoke pollution of towns. Israel Program for Scientific Translations, Jerusalem, 202. (From original Russian "Zadymlenie gorodov" Izdatel'stvo Ministerstva Kommulnogo Khozyaistva RSFSR, Moskva-Leningrad, 1949).

Smith, M. (ed.) (1968). Recommended guide for the prediction of the dispersion of airborne effluents. *Am. Soc. Mechan. Eng. Pap.,* 85 pp. New York.

Terpitz, W. (1965). Der Einfluss des Stadtdunstes auf die Globalstrahlung in Köln. Dissertation, Universität Köln, 103 pp.

Tomasi, C., Guzzi, R., and Vittori, O. (1975). The "SO_2–NH_3-solution droplets" system in an urban atmosphere. *J. Atmos. Sci.* **32**, 1580–1586.

Tuesday, C. S. (ed.) (1971). "Chemical Reactions in Urban Atmospheres," 283 pp. American Elsevier, New York.

4

Urban Energy Fluxes

4.1 SOLAR RADIATION

Little imagination is needed to gauge the effect of urban pollutants on the incoming solar radiation, especially the particulates, which scatter and absorb the sun's rays. It has long been known that urban areas have less sunshine than their surroundings. This is an urban effect that can not be doubted because sunshine duration is governed by the general weather situation so that differences in small areas are induced by local effects. Only in mountainous regions are similar mesoclimatic effects noted. In industrial cities the loss in sunshine duration can be between 10 and 20 percent. Similar losses are observed in terms of the energy received at the surface below the urban dust shield. Hufty (1970) noted that in Liège, Belgium, on days with high pollution the city lost 55 minutes of sunshine per day compared with surroundings. In London the inner city has 16 percent less sunshine duration, and the suburbs of the city have 5 percent less than the countryside (Chandler, 1965).

The urban reduction in energy received at the ground is greatest at low solar elevations when the relative thickness of the turbid layers is greatest. The losses are less at high solar elevations. As early as 1934, Steinhauser published some representative data for a few Central European cities, as shown in Table 4.1.

The losses are greatest at the low solar elevations, i.e., in the early morning and late afternoon hours. In winter and autumn the frequent low-level inversions contribute to the accumulation of pollutants and hence to the radiation loss. In spring the generally higher wind velocities and, in summer, the greater convection contribute to the dispersal of pollutants and thus the relatively smaller radiation loss.

Similar results were obtained by Terpitz (1965), who compared the solar energy received on a horizontal surface, the so-called global radiation, in the rather clean air at Trier with the polluted air at Cologne in West Germany. Although the two places are 130 km (81 miles) apart they have otherwise very comparable weather conditions. The radiative reduction was expressed in terms of the percentage of the energy that would be expected in a pure atmosphere where only molecular scattering (Rayleigh scatter) takes place. The mean monthly values of reduction are shown in Table 4.2. In this case, too, the cold season is the time when the metropolis receives much less radiation than the small town. In summer the difference nearly vanishes, but in December it reaches 25 percent of theoretically expected radiation in a clean atmosphere.

There are numerous comparable measurements from other localities. The observations show an annual value of 18 percent for Boston that is nearly identical to that for Cologne (Hand, 1949). For Canadian stations the loss is less: 9 percent for Montreal and 7 per-

TABLE 4.1

Percent Loss of Solar Radiation at Urban Sites Compared to Surrounding, Less-Polluted Countryside[a]

Solar elevation (deg)	Equivalent optical air mass (90° = 1)	Season			
		Winter	Spring	Summer	Autumn
10	5.4	36	29	29	34
20	2.9	26	20	21	23
30	2.0	21	15	18	19
40	1.6	—	15	14	16

[a] Adapted from Steinhauser (1934).

TABLE 4.2

Percent Reduction of Radiation on a Horizontal Surface at Trier and Cologne Compared with Alteration by Rayleigh Scattering[a]

Month	Trier	Cologne	Month	Trier	Cologne
I	70	50	VII	70	67
II	67	44	VIII	74	68
III	71	57	IX	65	61
IV	76	65	X	72	66
V	74	69	XI	60	40
VI	70	69	XII	71	46
Year:	70	59			

[a] From Terpitz (1965).

cent for Toronto (East, 1968). Japanese industrialized cities show rather high radiation losses. Nishizawa and Yamashita (1967) cite a range of 12–30 percent loss for Tokyo compared to a neighboring site.

In contrast, for a 19-month interval in St. Louis, Peterson and Stoffel (1980) found considerably less solar radiation depletion for all wavelengths. Their findings are shown in Table 4.3.

In an intensive survey of conditions in the British Isles, Unsworth and Monteith (1972) concluded that on an average, the direct beam of the sun is depleted by 38 percent. However, the diffuse radiation flux from the sky is greatly increased. When compared with diffuse radiation at the cleanest site in Britain in polar air, the average urban conditions show a 235 percent increase. The total radiation flux in the urban areas is only 82 percent of the lowest value found in the rural areas.

TABLE 4.3

Irradiation Depletion in St. Louis[a]

Site	Summer	Winter	Annual
Urban	2	4.5	3
Suburban	1	2	1.5

[a] In percent of rural area for cloudless sky conditions. (After Peterson and Stoffel, 1980.)

The attenuation is not evenly distributed over the solar spectrum. The shorter wavelengths are particularly affected. The ultraviolet suffers far more loss than the infrared. In many industrial cities in winter nearly all energy in wavelengths below $\lambda = 400$ nm is completely absorbed. There are not nearly as many spectral observations available as for total radiation, but an apparently representative example is available for Paris (Maurain, 1947) as shown in Table 4.4. The reduction of the ultraviolet by a factor of 10 is the most notable result. In the visible and the infrared the reduction is only 7 and 4 percent, respectively.

Measurements made in Leipzig and surroundings showed that in the urban area the heat portion of the spectrum has a considerably larger share of the radiation that reaches the ground than in the countryside (Bielich, 1933). For low solar elevations ($<25°$) the spectral distribution showed 6 percent more in the infrared of the total energy than in the nearby rural area. At high solar elevation ($>40°$) the difference diminished, but the relative share of the long wavelengths was still 3 percent higher in the city than in the country.

In the visible part of the spectrum, often referred to as illumination, the reduction is quite different in various cities. Steinhauser *et al.* (1955) pointed out that there is considerable compensation for the loss in the direct radiation beam by diffuse sky light. Nonetheless, these authors estimate that for Vienna, Austria, the illumination loss in summer is 10 percent and about 18 percent in winter, compared with the countryside.

The formidable loss of ultraviolet radiation in the direct beam has been frequently observed. In Leicester, England, a town of 210,000, winter values in the 300-nm band showed 30 percent less than the

TABLE 4.4

Spectral Partition[a] of Solar Radiation In and Near Paris[b]

Spectral region	Paris center	Outskirts
Ultraviolet	0.3	3.0
Extreme violet	2.5	5.0
Visible	43	40
Infrared	54	52

[a] In percent of total intensity.
[b] After Maurain (1947).

countryside (Department of Scientific and Industrial Research, 1947). In the Los Angeles Basin, measurements on individual days indicated ultraviolet irradiance losses of 50 percent, and over a period of sustained measurements from August to November, 1973, an average reduction of 11 percent at the International Airport and 20 percent at El Monte (Peterson *et al.*, 1978). At the same time total radiation was only attenuated by 6 percent at the airport and 8 percent at El Monte. The passing of pollutant plumes can be distinctly noted in ultraviolet radiation at Riverside. When such a plume from Los Angeles moves to the far west side of the Basin, reductions in ultraviolet irradiance of 25 percent can be observed at midday (Pitts *et al.*, 1968).

In growing urban areas very definite trends in solar radiation are now documented. An interesting case is the Tel-Aviv, Israel, area, which grew from 350,000 inhabitants in 1946 to 1.2 million in 1974. A series of global radiation records for the decade 1964–1973 at Bet-Dagan, 10 km to the southeast, in the suburbs, showed for summer an overall decrease of 3 percent and a 7 percent decrease for cloudless days (Fig. 4.1). This is an indication of formation of photochemical smog (Manes *et al.*, 1975).

There are some welcome indications that these effects are reversible. Pollution abatement is showing some results. An example is the central part of London, where the creation of a smoke-free zone after the air pollution disaster of December, 1952, is showing gratifying changes. Open fireplaces and the use of coal were replaced by central heating and gas. A comparison of the "clean" decade 1958–1967 with the "dirty" era (from 1931 to 1957) showed a no-

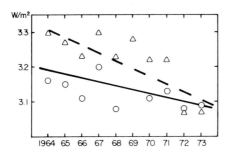

Fig. 4.1 Time series of global radiation at Bet-Dagan, Israel: dashed line (triangles) refers to clear days only; solid line refers to daily averages (adapted from Manes *et al.*, 1975).

table improvement in sunshine duration, especially in the heating season, as shown in Table 4.5 (Jenkins, 1970).

A very useful measure of atmospheric suspensions is the turbidity factor, which can be expressed as

$$T = \frac{\ln I_0 - \ln I - \ln S}{m E_R \ln e} \tag{4.1}$$

where

I_0 solar radiation extraterrestrially (solar constant)
I solar radiation at the earth's surface
S correction factor of the seasonably changing distance of sun–earth
m optical air mass
E_R extinction factor for pure dry air (Rayleigh)
e water vapor

In abbreviated terms one can describe this measure as a ratio of the prevailing attenuation of solar radiation to that produced by a clean atmosphere where only the molecular extinction (Rayleigh scatter and absorption) occurs. A correction for water vapor must be made because of its absorption in the infrared. For this reason an analogous turbidity factor T_K for the shorter wavelengths <625 nm is more useful in urban areas (Dogniaux, 1970).

Again, some interesting data have emerged from Tel-Aviv, where an early period of data from 1930 to 1934 is available. At that time T_K was 1.414; it had increased during 1961–1968 to 1.537 or about an 8 percent urbanization increase in turbidity (Joseph and Manes, 1971). A particularly striking example of the distribution of turbidity over a metropolitan area was published by Steinhauser (1934), Fig. 4.2. He depicted the total turbidity factor in Vienna, Austria, on a typical winter stagnation day. During that season water vapor was not as much of a contributing factor as in summer. A light SE wind

TABLE 4.5

Percentage Increase in Cold-Season Sunshine in London for the Interval 1958–1967 over the Prior Long-Term Average[a]

Month	IX	X	XI	XII	I	II	III
Percent increase	15	26	40	72	55	16	17

[a] After Jenkins (1970).

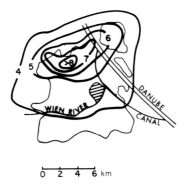

0 2 4 6 km

Fig. 4.2 Distribution of the turbidity factor on a stagnation day in Vienna, Austria (after Steinhauser, 1934).

had blown the pollution plume to the NW of the most densely built-up area of the city. The lowest values of T ($= 3.5$) were found upwind at the edge of urban area. The highest value was $T = 8.9$, a 250 percent increase. In general, values over 4 are considered as indicating notably polluted air.

4.2 OTHER FLUX PARAMETERS

The interaction of solar, atmospheric, and terrestrial radiation at the earth's surface without any complicating anthropogenic factors is itself a very complex phenomenon. Add the man-made changes and it becomes a formidable problem. The basic theory looks deceptively simple:

$$\pm Q_N = Q_I(1 - A) + Q_{L\downarrow} - Q_{L\uparrow}$$
$$= \pm Q_S \pm Q_H \pm Q_E + Q_P \qquad (4.2)$$

[at night the term $Q_I(1 - A)$ vanishes], where

Q_N net energy balance
Q_I incoming short-wave radiation received at the surface (both direct and diffuse)
A albedo of the surface (reflectivity)
$Q_{L\downarrow}$ long-wave atmospheric radiation downward

$Q_{L\uparrow}$ long-wave radiation upward emitted by the surface
Q_S heat flux into and out of the ground or other surfaces
Q_H sensible heat transfer between atmosphere and ground
Q_E heat loss by evaporation from surface (or plant cover) or gain by condensation (dew or frost formation)
Q_P heat production or heat rejection from man-made sources, including human and animal metabolism

Each one of these factors is different in urban areas from in the countryside and deserves special discussion. Q_I was already reviewed in the previous section. It is the easiest to deal with because it is readily amenable to measurement by pyranometers, and although adequate data sources in urban areas are as yet scarce, the trend toward solar energy use will encourage more of these measurements in the future. Q_N, the net energy gain or loss can also be directly measured, although here, too, the available data are generally restricted to very short intervals of observation.

The value of A, the albedo, has been measured on a number of occasions from aircraft, usually covering both urban and rural areas. Kung *et al.* (1964) first showed the differences that existed in the various environments and the notable seasonal variations induced in higher latitudes by snow cover. These early values indicated 10–30 percent lower values for urban than for rural albedos. More recently, in an air pollution study in the St. Louis area, Dabberdt and Davis (1974) gave the values shown in Table 4.6 for various land uses, obtained during a summertime flight at an altitude of 160 m.

There is about a 4 percent difference between distinctly rural and urban areas. Even though this may seem quite small, it is a significant factor when one deals with the large short-wave energy income from sun and sky. These summer values found close corroboration from another summer study in the St. Louis METROMEX project by White *et al.* (1978), who found albedos of 16.5 percent for the agricultural rural areas and 11.5 percent for the compact residential sectors of the city. Measurements on individual building materials, road surfaces, and parking lots showed albedo values as low as 5 percent. On the other hand, parkland in full leaf may reflect as much as 20 or 25 percent of the incoming short-wave radiation.

Differences in the component Q_L for urban and rural areas have not been exactly determined by observations. In the few cases where Q_L has been separately determined during the daytime simul-

TABLE 4.6

Albedo Values Measured 160 m above the Surface in the St. Louis Area on a Flight of August 9, 1979[a]

Land use	Average albedo (%)
Farmland	14.7
road, some trees	15.4
Woods	
some fields, roads	16.6
fields, some farmland	16.5
Mostly woods, some fields	16.2
New suburban housing tract	16.6
Old urban residential	12.2
commercial	12.1
Commercial, industrial, old housing	13.8

[a] After Dabberdt and Davis (1974).

taneously, the values in these different environments seem to be closely alike. Theoretically, the difference in radiation-absorbing aerosols and gases, such as particles and increased concentrations of CO_2 and other absorbing gases, leads one to expect a somewhat higher $Q_{L\downarrow}$ component in the urban than in the rural area (Ackerman, 1977). Calculations lead to a value of not more than 1–1.5 percent difference in this flux in the two environments. One set of measurements in the Tokyo area (Aida and Yaji, 1979) yielded, however, for nocturnal values an average of 5.7 percent difference, with a range of about 1–10 percent (for 14 measured values). The cause for the increase is the higher temperature of the aerosol layer above the city.

The absorption and scattering of radiation by haze layers had been observed in the London, England, area two decades ago by Roach (1961). Measurements were made from an aircraft in the spectral range between 300 and 3000 nm. There was considerable attenuation of the solar radiation. This included a 5 percent backscatter of the total radiation, 5 percent absorption in the visible, 15 percent in the long waves, and the remainder was a strong forward scatter. This attenuation could lead to heating rates of 5°C/day in the lowest haze layer. Such warming would lead to a different vertical temperature profile over the city than in the country, with an isothermal layer.

Using an empirical model of Nakagawa (1977), Aida and Yaji (1979) calculated the downward flux from aerological observations:

$$Q_\downarrow = \sigma T^4[0.127 + (-0.114\Gamma^3 - 0.168\Gamma^2 - 0.173\Gamma + 0.603)$$

$$\times(0.0000438e^3 - 0.001e^2 + 0.123e + 0.05)^{0.107}] \qquad (4.3)$$

where

$Q_\downarrow$ downward flux (W m^{-2})
σ Stefan–Boltzmann constant [$5.6696K \times 10^{-8}$ W m^{-2} K^{-4}]
T surface temperature (K)
e surface water vapor pressure (mbar)
Γ vertical temperature gradient (K mbar^{-1})

The calculations gave a 4 percent increase in downward flux over the urban area based on temperature structure alone. These different temperature profiles over the rural and urban areas have been variously simulated, and the role of the polluted layer has been explored (Atwater, 1971; Bergstrom and Viskanta, 1973; Viskanta *et al.*, 1977; Viskanta and Daniel, 1980). All these studies confirm that the pollutant layer plays a major role in the urban energy budget.

The factor $Q_{L\uparrow}$ is more amenable to measurement as a separate flux. A pyrgeometer will directly measure this element, but most of the time the value of this component is determined from the surface temperature

$$Q_{L\uparrow} = \epsilon\sigma T^4 \qquad (4.4)$$

where

ϵ emissivity
σ Stefan–Boltzmann constant
T absolute temperature (K) of the surface

There are a large number of surface-temperature measurements available. In earlier years this was determined by thermocouples. Presently these observations are made by infrared thermometry either from aircraft near the surface or from satellites. The first helicopter observations of this element were made by Lorenz (1962). He clearly documented the fact that in daytime with sunshine the surface temperatures of water, forest, and farmland stayed cool, and even small settlements showed higher values. Paved areas, such as runways, were always warmer than the surroundings. In midday an asphalt street was 17.9°C hotter than the air, a hangar roof + 17.4°C, a taxistrip + 14.1°C, and a small village + 3°C. Measurements of Kessler (1971) in Bonn, West Germany, yielded for an asphalt street the maximum temperature 23.5°C above air temperature and minimum temperature + 2.6°C. Over a grass surface the corresponding values were: maximum, + 9.4°C, minimum, − 2.9°C.

It must be emphasized that in clear weather, the surface temperature is nearly always different from the air temperature, measured in a meteorological shelter, both day and night. In overcast, windy conditions the two temperatures may coincide. On a cloudless day in summer the contrasts show up at an early hour. Table 4.7 shows a set of measurements from a survey of the new town of Columbia, Maryland. These data were taken from a helicopter at elevations of 50–100 m above the surface, with an air (shelter) temperature of 29–30°C during the flight, which took place about 1½ hr before solar noon on a summer day (Landsberg, 1969). From the same study of changes brought about by urbanization, two typical cases of day and night conditions under clear skys are shown in Table 4.8.

Other nocturnal observations were reported by Vilkner (1962) who noted a 12°C difference between moist pastures outside the town of Greifswald, East Germany, and the densely built-up center on a clear night. Hence nocturnal differences are of great influence on the vegetation, but all accounts agree on even greater daytime differences. In Columbia, Maryland, a difference of as much as 26°C between the surface temperature on a parking lot and the air temperature was observed. In that growing town it was possible to relate the maximum midday surface temperature differences between the

TABLE 4.7

Infrared Measurements of Surface Temperatures on a Sunny Morning in Columbia, Maryland

Land use	Temperature (°C)
Lake	27.5
Forest	27.5
Farmland	30.8
Parkland	31.0
Open housing areas	32.2
Built-up spaces	34.7
Parking lots and shopping center	36.0

TABLE 4.8

Surface Temperatures and Difference in Air Temperature in the Developing Urban Area of Columbia, Maryland

Type of surface	Day		Night	
	Surface temp. (°C)	Air temp. difference (°C)	Surface temp. (°C)	Air temp. difference (°C)
Lake	26	−1	12	0
Bare soil	35	+4	6	−7
Grass	30	+3	2	−10
Asphalt	41	+14	12	−2

urban and rural areas to the building density, as shown in Fig. 4.3. The extremely high temperatures that can be reached in the afternoon hours of a sunny day have been noted in Vienna, Austria, by Steinhauser *et al.* (1959). For an August day with air temperatures ranging in the city from 22 to 32°C, these authors reported street surface temperatures of 25°C in the shade and up to 51°C in the sun. Metal-roof temperatures of 60°C under such conditions are not uncommon.

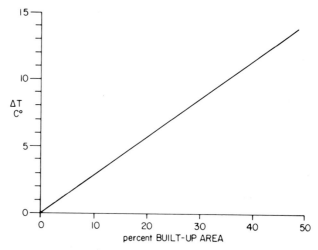

Fig. 4.3 Midday surface temperature excess (urban−rural, °C) on sunny days as a function of the percentage of built-up area.

A great deal of information on surface temperatures can be obtained from infrared scans of satellites. Only in recent times has it been possible to move from qualitative information to quantitative data (Matson and Legeckis, 1980). The resolution is still rather coarse but the pictures obtained are impressive, especially at night (Fig. 4.4). Some analyses have been performed for daytime values. An example is shown in Fig. 4.5 for the Baltimore–Washington corridor.

Such observations enable one to calculate the outgoing radiative energy flux ($Q_{L\uparrow} = \epsilon\sigma T^4$). The emissivity of urban surfaces can be set at 0.96, although this is often overlooked and the factor is

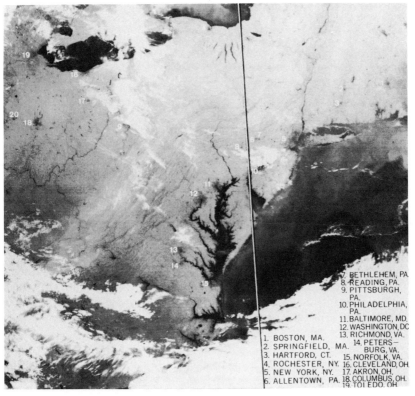

7. BETHLEHEM, PA.
8. READING, PA.
9. PITTSBURGH, PA.
10. PHILADELPHIA, PA.
11. BALTIMORE, MD.
12. WASHINGTON, DC
13. RICHMOND, VA.
14. PETERSBURG, VA.
15. NORFOLK, VA.
16. CLEVELAND, OH.
17. AKRON, OH.
18. COLUMBUS, OH.
19. TOLEDO. OH

1. BOSTON, MA.
2. SPRINGFIELD, MA.
3. HARTFORD, CT.
4. ROCHESTER, NY.
5. NEW YORK, NY.
6. ALLENTOWN, PA.

Fig. 4.4 Satellite picture of urban nocturnal heat islands by infrared sensing; dark areas are warm. (Courtesy of National Environmental Satellite Service, National Oceanographic and Atmospheric Administration.)

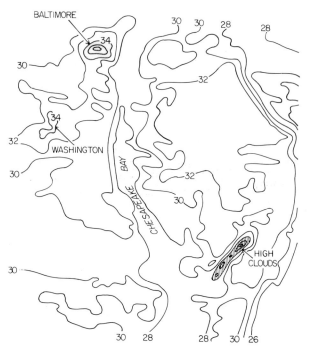

Fig. 4.5 Isotherms of surface temperatures (OC) interpreted from satellite infrared sensor transmissions. (Adapted from National Environmental Satellite Service, National Oceanographic and Atmospheric Administration interpretation.). Note the distinct heat islands of Washington, D. C., and Baltimore.

neglected. In actuality, this makes little difference in view of the fact that both in daytime and at night, as noted above, there is such a multiplicity of surface temperatures. These enter the equation as 4th powers of absolute values. Then a 10° nocturnal difference from, say, 278 to 288 K raises the outgoing energy by 14 percent. Thus it is quite difficult to arrive at a value for $Q_{L\uparrow}$. Fuggle and Oke (1970) remarked: "It appears that the major obstacle to the implementation of the heat balance approach to the urban heat island, and other urban climatic phenomena, is connected with the almost overwhelming diversity and irregularity of urban surfaces. Each one of these surfaces possesses a different albedo, emissivity and heat capacity."

Oke and collaborators have tried to overcome the difficulties by a series of studies investigating the infrared fluxes in urban areas (Oke and Fuggle, 1972; Fuggle and Oke, 1976; Nunez and Oke, 1976; Nunez and Oke, 1977). They introduced the very useful distinction of radiation above the roof level and from what they called the "urban canyon." This term refers to the streets and walls of houses and buildings. These are considered as the essential feature of an urban surface in contrast to the usually flat surfaces of the countryside. From measurements in Montreal these authors obtained as an average of 12 clear nights the values for the radiation balance shown in Table 4.9.

The differences are relatively small. Increased atmospheric long-wave radiation downward over the city is somewhat exceeded by outgoing radiation from the surface. The calculated urban radiative cooling rate above the roof level was determined by Fuggle and Oke (1976) by measuring Q_N and comparing it with actual temperature measurements. The radiative cooling was obtained from the relation

$$\left(\frac{\Delta T}{\Delta t}\right)_{rad} = \frac{\text{div } Q_N}{\rho c_p} \tag{4.5}$$

where

T air temperature
t time
Q_N net radiation
ρ air density
c_p specific heat at constant pressure

TABLE 4.9

Urban and Rural Long-Wave Radiation Components in the Montreal Area[a,b,c]

	$Q_{L\downarrow}$	$Q_{L\uparrow}$	Q_N
Urban	31.3	−40.1	−8.8
Rural	29.8	−38.2	−8.4
Δ_{u-r}	1.5	1.9	+0.4

[a] Oke and Fuggle (1972).
[b] All values in mW cm^{-2}.
[c] Negative sign indicates heat loss.

The results showed that the calculated radiative rate is always larger than the observed cooling rate. The difference between the two rates was about 3°C hr^{-1}. The explanation is that the radiative cooling is compensated by sensible heat flux convergence.

The complexity of the "urban canyon" energy fluxes are immediately obvious by a glance at the schematic representation of Fig. 4.6 (Nunez and Oke, 1977). A multiple-layer scheme for evaluation was adopted and actual measurements were made in an alley 7.54-m wide, a west wall 5.59-m high, and an east wall 7.31-m high. The energy balance of the ith canyon level is given by:

$$Q_{N_i} = Q_{H_i} + Q_{E_i} + Q_{S_i} \tag{4.6}$$

using the same designators as in Eq. (4.2).

The principal conclusions from the experiment are that in daytime Q_H is the principal element carrying heat from the canyon, representing 64 percent of the net radiation. Clearly, wind flow through or across the canyon must be important. Most of the remainder of the energy is stored in the ground and walls. At night Q_H is small because the data were gathered when winds are weak. The radiative flux divergence is large and the net loss is partially compensated for by heat stored in the walls and street during the day. Q_E under the circumstances is small.

In the Columbia, Maryland, experiment (Landsberg, 1973), a complete set of measurements of the elements of the radiation balance on a clear day near equinox, with very light winds (<3 msec^{-1}), was made. The results are shown graphically in Fig. 4.7. In the rural area, in daytime, the evaporative component Q_E is quite large and almost negligible in the urban area. The reflected short-wave radiation because of the high albedo in actively growing vegetation is also

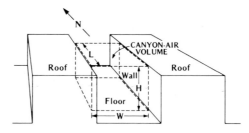

Fig. 4.6 The "urban canyon" in schematic representation (from Nunez and Oke, 1977).

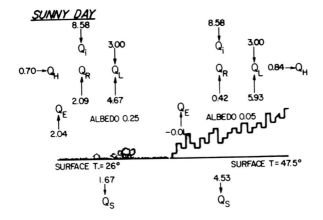

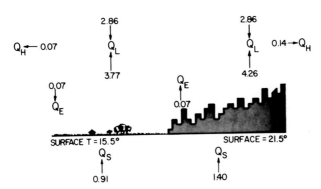

Fig. 4.7 Heat flux data from Columbia, Maryland, experiment. (All values given in W m^{-2}.)

large in the rural area and small in the urban area. The outgoing long-wave radiation is larger in the city than in the rural area, but the greatest difference between them is the stored heat Q_S, which is moderate in the vegetated rural area (19 percent) but to over 50 in the city of the incoming radiation. This is well reflected in the high surface temperature. The nocturnal values are not as startlingly different between the two environments.

Some new information on $Q_{L\downarrow}$ has become available from interpretations of data gathered by an Explorer satellite, launched as Heat

Capacity Mapping Mission (HCMM) by NASA's Goddard Space Flight Center. Surface temperatures were calculated from emissions in the infrared (10.5–12.5 μm). The HCMM satellite flew at 620-km altitudes and had a horizontal resolution of 500 m. Data obtained on a nearly clear summer day (June 6, 1978) at the time close to maximum solar elevation (1300 hr EST) over parts of the heavily urbanized northeastern United States are shown in Table 4.10 (Price, 1979). This table shows the radiative temperature excesses of the urban areas above the surrounding countryside and the calculated radiated power.

There is an obvious, if loose, relation of the excess temperatures to the population and extent of the urban areas. However, we deal with an isolated observation, and for very systematic linkages one would have to resolve the urban fabric into finer segments. In fact, in some of the cities isolated points with additional temperature excesses of 2–3°C were noted, possibly attributable to power plants

TABLE 4.10

Surface Temperature Differences and Excess Radiated Power[a] for Urban Areas in the Early Afternoon in Summer[b,c]

City	Metropolitan population ($\times 10^3$)	Area (km²)	Temp. diff. (°C)	Radiated power (kW)
New York City	7895	547	17	40,000
Providence, RI	720	48.2	13.2	3190
Hartford, CT	817	26.2	15.0	1770
Schenectady, NY	78	5.3	15.0	368
Bridgeport, CT	157	14.2	12.1	954
Syracuse, NY	197	6.5	10.9	417
Binghampton, NY	64	6.0	12.4	394
New Haven, CT	745	5.8	11.2	372
Worcester, MA	637	5.1	11.5	327
Albany, NY	116	3.7	10.3	227
Stanford, CT	109	3.2	11.2	202
Waterbury, CT	108	1.2	10.9	74
Fitchburg, MA	43	1.2	11.2	131
Troy, NY	62	1.4	10.3	86
Pittsfield, MA	57	0.7	9.7	42

[a] 9°C was chosen as threshold for calculating the excess radiation. Some smaller localities from the original tables were omitted.
[b] Northeastern United States.
[c] From Price (1979).

and industrial agglomerations. Although "ground truth" was not ascertained the figures look entirely reasonable in the light of the measurements discussed earlier.

4.3 MAN-MADE FACTORS IN THE URBAN HEAT BALANCE

Radiation conditions, especially in the canyon streets of the inner cities, are greatly complicated by the change in horizon, which affects the duration of sunshine and illumination. There are also radiative interactions between building fronts in narrow streets and between the buildings and the street surface. This interaction is determined by the height of houses or buildings z_b, the width of the street w_s, and the azimuth of the street ϵ. One can then define an index of street narrowness N as

$$N = \frac{z_b}{w_s} \tag{4.7}$$

This ratio also defines an angle σ of the horizon created by the building line so that

$$\tan \sigma = N$$

The value of N for various values of z_b and w_s are shown in Table 4.11 (Kaempfert, 1949).

The relation between the azimuth angle of the street ϵ, N, and the horizon angle σ is shown in Table 4.12 adopting the convention that

TABLE 4.11

Index of Street Narrowness

z_b (m)	w_s (m) 5	10	15	20
5	1	0.5	0.33	0.25
10	2	1	0.67	0.5
15	3	1.5	1	0.75
20	4	2.0	1.33	1

TABLE 4.12

Dependence of Horizon Angle σ on Index of Street Narrowness N and Street Azimuth

			N		
ϵ°	0.2	0.5	1.0	2.0	5.0
0	11.3	26.0	45.0	63.4	78.7
20	10.7	25.2	43.2	62.0	78.0
40	8.7	21.0	37.5	56.8	75.4
60	5.7	14.0	26.6	45.0	68.2
80	2.0	5.0	9.9	19.3	41.0

$\epsilon = 0$ points south and that westerly azimuths are positive and easterly azimuths are negative (only the positive values for σ are given in Table 4.12). This information enables one to superimpose on the theoretical march of the sun above the horizon the restrictions introduced by the building contours. For architectural purposes the horizon can also be constructed empirically by whole-sky cameras (Pleijel, 1954). The artificial horizon will retard sunrise and advance sunset for the street dwellers and thus reduce the available solar radiation and illumination to the inhabitants.

The attenuation of sky illumination and radiation by buildings on the opposite side of the street can be gauged from Table 4.13 where the percentage loss is related to the depth below the roof line in

TABLE 4.13

Illumination Loss Caused by Urban Street Width and Building Height

Ratio of depth below roof/w_s	Illumination loss (%)
0	50
0.25	58
0.5	65
1	75
2	85
3	90
4	92

TABLE 4.14

Air and Surface Temperatures (°C) in a Courtyard on a Sunny Day

| | | | | Surface | | | | |
| | | | | | Walls facing | | | |
Weather conditions	Time	Air	Grass	Courtyard	N	E	S	W
3/10 clouds, full sunshine, wind 3 m sec⁻¹	1620	30.6	33	50	32	35	35	50
Sunset, wind 1 m sec⁻¹	1934	28.3	29	33	31	31	32	34
Clear, calm	2115	25.6	23	31	28	28	30	30

terms of the width of the street. It might be noted here in passing that illumination in cities is always less than in the rural environs because of the interception of light by pollutants. Only a few systematic sets of measurements are available. For the steel-manufacturing town of Zaporozhe in the Ukraine, Fedorov (1958) showed that at the highest solar elevations in June illumination in town was reduced by 5 percent and in December when the sun was lowest by 13 percent.

Measurements of infrared temperatures on walls and surfaces in courtyards show the multiplicity of micrometeorological interactions that must be expected in the urban radiative processes. Table 4.14 shows a set of measurements on a bright summer day in a courtyard (32 × 42 m) of buildings 18-m high, with grass surfaces on the outside (Landsberg, 1970). The startling contrasts that remain in a small space, even after the direct radiation ceases, show that intricate internal heat fluxes will rule the energy balances over large portions of a city.

The wall and ground temperatures also govern the energy exchange with the interior of buildings. This is a reciprocal flux that is of great importance for the design of buildings for optimal energy efficiency. Sagara and Horie (1978) have explored this problem in Senri New Town in Japan. The radiative model employed follows the classical radiation equations for wall surfaces. The daytime solar flux and sky radiation affect the walls, as follows (Kondratyev, 1977):

Vertical surface facing $\begin{cases} \text{South,} & Q_{I_S} = Q_\perp \sin h \cos \psi \\ \text{East or west,} & Q_{I_{E(W)}} = Q_\perp \cos h \sin \psi \\ \text{North,} & Q_{I_N} = Q_\perp [\sin \delta \cos \phi \\ & \quad - \cos \delta \cos \phi \cos \Omega] \end{cases}$ (4.8)

where

Q_I direct solar beam energy hitting wall
$Q_\perp$ solar energy received at earth's surface normal to sun's rays
h solar elevation above horizon
δ solar declination
ϕ latitude
ψ solar azimuth
Ω solar hour angle

For the sky radiation we have:

$$Q_{Sk} = I_0 \frac{\sin h}{2} \left[\frac{1 - \tau \operatorname{cosec} h}{1 - 1.4 \ln \tau} \right]$$ (4.9)

where

Q_{Sk} diffuse sky radiation
I_0 solar radiation at boundary of atmosphere
τ transparency

The energy loss from the surface of the walls is, according to Sagara and Horie (1978), governed by

$$Q_{wo} = \sum_{i=1}^{n} \epsilon_w \epsilon_i \sigma \left[\left(\frac{T_w}{100} \right)^4 - \left(\frac{T_i}{100} \right)^4 \right] \eta$$
$$+ \epsilon_w \sigma \left[\left(\frac{T_w}{100} \right)^4 - \left(\frac{T_a}{100} \right) (a + b\sqrt{f}) \right] \left(1 - k \frac{M}{10} \right) \eta \quad (4.10)$$

where

Q_{wo} outgoing long-wave radiation from the walls
T_w, T_i, T_a absolute temperature of: outer wall, inner wall, and air near wall, respectively
ϵ_w, ϵ_i wall emissivities, outer and inner wall
σ Stefan–Boltzmann constant
M cloud amount

k constant depending on cloud height

a, b constants

η sky aspect factor of wall

Some measured values for summer and winter in the Japanese town are shown in Fig. 4.8. In summer the notable gains of the east wall in the early morning hours are remarkable. The nocturnal losses are not markedly different for the various wall exposures. In winter east and south gain most heat in the morning hours. Nocturnal heat losses are highest for the east wall and lowest for the south wall, but the short measuring interval of only three days permits no categorical statements. The presence of balconies on the building complex produces further complications.

An important quantity in the urban heat balance is obviously the energy used in the city and its ultimate release into the atmosphere.

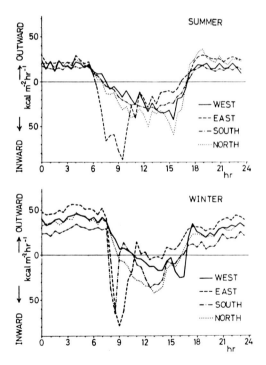

Fig. 4.8 Heat fluxes through external wall surfaces in summer and winter (after Sagara and Horic, 1978).

Only a limited store of information on this flux is available; estimates have been made for a few urban areas. The problem lies with the statistics on fuel use. Not all fuel delivered to a given city is used there, such as gasoline sold there but burned by driving in the countryside. Often the statistical districts encompass substantial amounts of rural area economically linked to a particular city.

For a few cities estimates can give an idea of what orders of magnitude are dealt with. Particularly notable are the differences of heat rejection by unit areas of various metropolitan localities as shown in Table 4.15.

The extraordinary concentration of energy use in the crowded quarters of Manhattan is not equaled elsewhere. Most other places cover a greater surface area and have considerably less population density. Only for a few cities are there more detailed studies for various subdivisions of metropolitan areas: London (McGoldrick, 1980), Montreal (East, 1971), and Sydney, Australia (Kalma *et al.*, 1973) are pertinent examples. The Australian study showed that the ratio of annual energy use in the inner city to the outlying sector of the Sydney Statistical Division was about 360 to 1. Such differences

TABLE 4.15

Metropolitan Energy Use[a]

City	Population (millions)	Area (km^2)	Energy flux density ($W\ m^{-2}$)
Manhattan	1.7	234	630
Moscow	6.4	878	127
Sydney (city)	0.1	24	57
Chicago	3.5	1800	53
Budapest	2.0	525	43
Aruba	0.06	180	35
Brussels	1.0	400	28
Cincinnati	0.54	200	26
West Berlin	2.3	233	21
Los Angeles	7	3500	21
Sheffield	0.5	49	20
Fairbanks	0.03	36	19
St. Louis	0.75	250	16
Hongkong	4.4	92	3

[a] Sources: Bach, 1970; Borisenkov, 1977; Kalma *et al.*, 1973; SMIC, 1971.

must find expression in the micrometeorological fabric of the urban area. Kalma *et al.* (1973) estimated that the annual artificial heat generated in Sydney, in 1970, was 24.74 J (234.5 × 10^{12} Btu). For Montreal, East calculated the anthropogenic heat production at 39 × 10^{16} J (9.33 × 10^{16} cal) or 24 percent of the total energy transformation of the urban area.

A very detailed map of energy use in Greater London for the year 1971 was undertaken by McGoldrick (1980). He found that the average daily artificial heat release in the outer districts is about 0–5 W m^{-2}. In the city center there are several square kilometers where the average exceeds 100 W m^{-2}, with a maximum for 1 km^2 in the inner city of 234 W m^{-2}. The daily global solar radiation for the area averages 106 W m^{-2}. It is estimated that Greater London rejects from domestic, commercial, institutional, industrial, and traffic energy use 17.4 GW/day during the year and 21.8 GW/day in December when energy use reaches its annual peak.

For one specific sector of use of energy a very good relation to meteorological elements exists, namely space heating. Turner (1968) showed for St. Louis that a very high correlation exists between daily temperatures in winter and the demand for gas and steam for heating purposes, as Fig. 4.9 demonstrates. The linear regression between temperature and energy use on weekdays explains 82 percent of the variance. East (1971) performed a similar analysis in which he related the space heating urban energy flux in Montreal to both temperature deviation from a specific base 65°F (18°C), the well-known degree day value, and a wind function. The linear regression had the following form

$$Q_{\mathrm{Sh}} = a + b(1 + u) \, \mathrm{dd} \qquad (4.11)$$

where

Q_{Sh} heat flux from space heating
a, b constants
u wind speed
dd degree-day value

Under very simplified conditions one can set the temperature rise attributable to the anthropogenic heat Q_{p} in urban areas to be

$$\Delta T_{\mathrm{Gp}} = \left[Q_{\mathrm{p}} a \, \frac{d\Theta}{dz} \bigg/ c_{\mathrm{p}} \phi \bar{u} \right]^{1/2} \qquad (4.12)$$

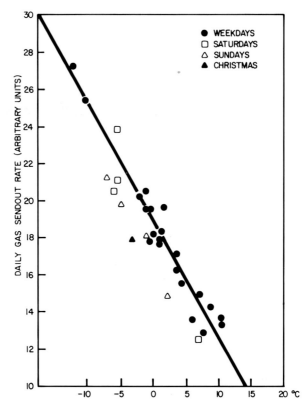

Fig. 4.9 Use of heating gas in St. Louis, Missouri, as a function of daily temperature (from Turner, 1968).

where

a area of the city

$d\Theta/dz$ vertical lapse rate of potential temperature Θ with height z

c_p specific heat of air at constant pressure

ϕ density of air

$\bar{u}$ mean wind speed

The fact that Q_p and $d\Theta/dz$ vary considerably in an urban area and that the lapse rate is itself a function of ΔT, makes this only a first approximation.

There has also been some discussion of the significance of meta-

TABLE 4.16

Anthropogenic Heat Production during the Course of the Day[a]

	Weighted share (%) at various daytimes				
	Hour				
Heat source	0800	1300	2000	Night	Daily share
Stationary	71	64	71	45	66.6
Mobile	69	45	25	12	33.3
Metabolism of organisms	0.05	0.2	0.1	0.02	0.1
$(W\ m^{-2})$ metabolic heat	0.36	0.29	0.26	0.14	0.26

[a] In relative values (Bach, 1970).

bolic heat of people and animals on the urban heat budget (Terjung, 1970). For a population of a million persons and a corresponding number of domestic animals one can estimate metabolic heat production at $\sim 5.3 \times 10^{15}$ J (5×10^{12} Btu). This, depending on the other anthropogenic heat production, is not more than maximally 3 or 4 percent of the total urban energy rejection and, in most cases, less than 1 percent. This component is usually neglected in urban energy budgets. Based on data collected in Cincinnati during summer, Bach (1970) estimated the contribution of various sources of anthropogenic heat production throughout the day, as shown in Table 4.16.

References

Ackerman, T. P. (1977). A model of the effect of aerosols on urban climates with particular applications to the Los Angeles Basin. *J. Atmos. Sci.* **34**, 531–547.

Aida, M., and Yaji, M. (1979). Observations of atmospheric downward radiation in the Tokyo area. *Boundary Layer Meteorol.* **16**, 453–465.

Atwater, M. A. (1971). The radiation budget for polluted layers of the urban environment. *J. Appl. Meteorol.* **10**, 205–214.

Bach, W. (1970). An urban circulation model. *Arch. Meteorol. Geophys. Bioclimatol. Ser. B.* **18**, 155–168.

Bergstrom, R. W., Jr., and Viskanta, R. (1973). Modeling the effects of gaseous and particulate pollutants in the urban atmosphere, Part I: Thermal structure. *J. Appl. Meteorol.* **12**, 901–912.

Bielich, F.-H. (1933). Einfluss der Groszstadtrübung auf Sicht und Sonnenstrahlung, 49 pp. Dissertation, Univ. Leipzig. Universitätsverlag von Robert Noske, Borna-Leipzig.

Borisenkov, Ye. P. (1977). Development of the fuel and energy base and its influence on weather and climate (translated title). *Meteorol. Gidrol.*, No. 2, 1–16.

Chandler, T. J. (1965). "The Climate of London," p. 122. (292 pp.) Hutchinson, London.

Dabberdt, W. F., and Davis, P. A. (1974). Determination of energetic characteristics of urban–rural surfaces in the greater St. Louis area. *Preprint, Symp. Atmos. Diffusion and Air Pollut., Santa Barbara,* pp. 133–141. Am. Meteorol. Soc., Boston.

Department of Scientific and Industrial Research (1947). Atmospheric pollution in Leicester: A scientific survey. *Atmos. Pollut. Res., Tech. Paper,* No. 1, 161 pp., London.

Dogniaux, R. (1970). Ambiance climatique et confort de l'habitat en site urbain: Aspects thermiques et lumineux. *In* "Urban Climates," *WMO Tech. Note,* No. 108, pp. 49–64.

East, C. (1968). Comparison du rayonnement solaire en ville et à la campagne. *Cah. Géographie de Québec* **12**, 81–89.

East, C. (1971). Chaleur urbaine à Montréal. *Atmosphere* **9**, 112–122.

Fuggle, R. F., and Oke, T. R. (1970). Infra-red flux divergence and the urban heat island. *In* "Urban Climates," *WMO Tech. Note,* No. 108, pp. 70–78.

Fuggle, R. F., and Oke, T. R. (1976). Long-wave radiative flux divergence and nocturnal cooling of the urban atmosphere, I. Above roof level. *Boundary Layer Meteorol.* **10**, 113–120.

Hufty, A. (1970). Les conditions de rayonnement en ville. *In* "Urban Climates," *WMO Tech. Note,* No. 108, pp. 65–69.

Jenkins, J. (1970). Increase in averages of sunshine in central London. *In* "Urban Climates," *WMO Tech. Note,* No. 108, pp. 292–294.

Joseph, J., and Manes, A. (1971). Secular and seasonal variation of atmospheric turbidity at Jerusalem. *J. Appl. Meteorol.* **10**, 453–462.

Kaempfert, W. (1949). Zur Frage der Besonnung enger Strassen. *Meteorolog. Rundsch.* **2**, 222–227.

Kalma, J. D., Aston, A. R., and Millington, R. J. (1973). Energy use in the Sydney area. *Proc. Ecolog. Soc. Australia* **7**, 125–142. *In* "The City as a Life System?" (H. A. Mix, ed.).

Kessler, A. (1971). Über den Tagesgang von Oberflächentemperaturen in der Bonner Innenstadt an einem sommerlichen Strahlungstag. *Erdkunde* **25**, 13–20.

Kondratyev, K. Ya. (1977). Radiation regime of inclined surfaces. *WMO Tech. Note,* No. 152, 82 pp.

Kung, E. C., Bryson, R. A., and Lenchow, D. H. (1964). Study of continental surface albedo on the basis of flight measurements. *Mon. Weather Rev.* **92**, 543–564.

Landsberg, H. E. (1969). Biometeorological aspects of urban climate. Tech. Note BN-620, 13 pp. Inst. for Fluid Dynamics and Appl. Math., Univ. of Maryland, College Park, Md.

Landsberg, H. E. (1970). Micrometeorological temperature differentiation through urbanization. *In* "Urban Climates," *WMO Tech. Note,* No. 108, pp. 129–136.

Landsberg, H. E. (1973). Climate of the urban biosphere. *In* "Biometeorology" (S. W. Tromp, W. H. Weihe, J. J. Bouma, eds.), Vol. 5, Pt. II, pp. 71–83.

Lorenz, D. (1962). Messungen der Bodenoberflächentemperatur vom Hubschrauber aus. *Ber. Deutsch. Wetterdienstes* **11** (82), 29 pp.

Manes, A., Goldreich, Y., Rindsberger, M., and Guetta, D. (1975). Inadvertment (sic!) modification of the solar radiation climate at Bet-Dagan. *Proc. Sci. Conf. Israel Ecol. Soc., Tel-Aviv, 6th*, pp. 224–232.

Matson, M., and Legeckis, R. V. (1980). Urban heat islands detected by satellite. *Bull. Am. Meteorol. Soc.* **61**, 212.

Maurain, C. H. (1947). "Le Climat Parisien," 163 pp. Presses Univ., Paris.

McGoldrick, B. (1980). Artificial heat release from Greater London, 1971. Physics Divisum Energy Workshop Rept. No. 20, 32 pp. Dept. of Physical Sciences, Sunderland Polytechnic, Sunderland.

Munn, R. E. (1973). Urban meteorology: Some selected topics. *Bull. Am. Meteorol. Soc.* **54**, 90–93.

Nakagawa, K. (1977). Atmospheric radiation from cloudless sky. *Geogr. Rev. Japan* **30**, 129–143.

Nishizawa, T., and Yamashita, S. (1967). On attenuation of the solar radiation in the largest cities. *Jap. Progr. Climatol., Tokyo*, 66–70.

Nunez, M., and Oke, T. R. (1976). Long-wave radiative flux divergence and nocturnal cooling of the urban atmosphere, II. Within an urban canyon. *Boundary Layer Meteorol.* **10**, 121–135.

Nunez, M., and Oke, T. R. (1977). The energy balance of an urban canyon. *J. Appl. Meteorol.* **16**, 11–19.

Oke, T. R., and Fuggle, R. F. (1972). Comparison of urban counter and net radiation at night. *Boundary Layer Meteorol.* **2**, 290–308.

Peterson, J. T., Flowers, E. C., and Rudisill, J. H. (1978). Urban–rural solar radiation and atmospheric turbidity measurements in the Los Angeles Basin. *J. Appl. Meteorol.* **17**, 1595–1609.

Peterson, J. T., and Stoffel, T. L. (1980). Analysis of urban–rural solar radiation data from St. Louis, Missouri. *J. Appl. Meteorol.* **19**, 275–283.

Pitts, J. N. Jr., Cowell, G. W., and Burley, D. R. (1968). Film actinometer for measurement of solar ultraviolet radiation intensities in urban atmospheres. *Environ. Sci. Technol.* **2**, 435–437.

Pleijel, G. (1954). "Computation of Natural Radiation in Architecture and Town Planning," 143 pp. Tech. Skrifter, Stockholm.

Price, J. C. (1979). Assessment of the urban heat island effect through the use of satellite data. *Mon. Weather Rev.* **107**, 1554–1557.

Roach, W. T. (1961). Some aircraft observations of fluxes of solar radiation in the atmosphere. *Q. J. Roy. Meteorol. Soc.* **87**, 346–363.

Sagara, K., and Horie, G. (1978). Effects of heat fluxes through external surfaces of the vertical walls on external thermal environment. *Jpn. Progr. Climatol.*, Japan Climatology Seminar (Tokyo), pp. 1–11.

SMIC (Rept. of Study of Man's Impact on Climate) (1971). "Inadvertent Climate Modification," 308 pp. MIT Press, Cambridge, Mass.

Steinhauser, F. (1934). Neue Untersuchungen der Temperaturverhältnisse von Grosstädten: Methode und Ergebnisse. *Bioklim. Beiblätter* **1**, 105–111.

Steinhauser, F., Eckel, O., and Sauberen, F. (1955). Klima und Bioklima von Wien. 1. Teil; *Wetter Leben,* Sonderheft 3, p. 17.

Steinhauser, F., Eckel, O., and Sauberer, F. (1959). Klima und Bioklima von Wien, III. Teil; *Wetter Leben* **11** (Sonderheft), 135 pp.

Terjung, W. H. (1970). Urban energy balance climatology: A preliminary investigation of the city–man system in downtown Los Angeles. *Geogr. Rev.* **60**, 31–53.

Terpitz, W. (1965). Der Einfluss des Stadtdunstes auf die Globalstrahlung in Köln. Dissertation, Univ. of Köln. 103 pp.

Turner, D. B. (1968). The diurnal and day-to-day variations of fuel usage for space heating in St. Louis, Missouri. *Atmos. Environ.* **2**, 339–351.

Unsworth, M. H., and Monteith, J. L. (1972). Aerosol and solar radiation in Britain. *Q. J. Roy. Meteorol. Soc.* **98**, 778–797.

Vilkner, H. (1961). Die Nachtemperatur am Erdboden in einer Stadt; *Z. Meteorol.* **15**, 141–147.

Viskanta, R., and Daniel, R. A. (1980). Radiative effects of elevated pollutant layers on temperature structure and dispersion in an urban atmosphere. *J. Appl. Meteorol.* **19**, 53–70.

Viskanta, R., Bergstrom, R. W., and Johnson, R. D. (1977). Radiative transfer in a polluted urban planetary boundary layer. *J. Atmos. Sci.* **34**, 1091–1103.

White, J. M., Eaton, F. D., and Auer, A. H., Jr. (1978). The net radiation budget of the St. Louis metropolitan area. *J. Appl. Meteorol.* **17**, 593–599.

5

The
Urban
Heat
Island

The most obvious climatic manifestation of urbanization is the trend toward higher air temperatures. This is also the theme most frequently discussed in the literature since its discovery by Luke Howard. It is present in every town and city. Several points, however, need emphasis. The differences that develop between an urbanizing area and the rural landscape greatly depend on the synoptic conditions. They are in essence a differentiation of topoclimates and as such depend on dissimilarity of radiative fluxes and turbulent exchanges. These contrasts are largest in clear, calm conditions. They tend to disappear in cloudy and windy weather. Thus the principal instrumentalities of a distinct urban climate are stationary high-pressure systems. Under these synoptic regimes temperature differences between urban and rural areas become large. Closed isotherms separate the city from the general temperature field, and this condition has become known as the *urban heat island*.[1] Its char-

[1] The term appears for the first time in the English-language meteorological literature in a paper by Gordon Manley (1958) in the *Quarterly Journal of the Royal Meteorological Society*, but it may have been used earlier elsewhere.

acter as a variable condition was best expressed by Linke: "The town climate is a good weather phenomenon which shows its greatest development in calm air and cloudless sky" (Linke, 1940).

5.1 DEVELOPMENT AND GROWTH OF THE HEAT ISLAND

The heat island is a reflection of the totality of microclimatic changes brought about by man-made alterations of the urban surface. Even a single building complex will show a different microclimate than an equal piece of land in its natural state. The paved surfaces and walls will store some of the heat received in daytime and give it off after sunset to its air environment. It has been noted in Section 4.3 how the diurnal flux of heat into and out of man-made surfaces proceeds. The natural order is further altered by the lack of evaporation in the cities. Solar energy that is used in the country in the morning to evaporate dew, guttation on plants, and frost is directly absorbed by the building materials. Evapotranspiration is also sharply reduced in the city because of the reduced plant cover. The rapid runoff after precipitation essentially eliminates water storage in and evaporation from soil.

The contrasts are most obvious in the late afternoon around and after sunset. A typical case was observed for a five-story brick building complex surrounding a paved court, 32 × 42 m. On the outside there was a wide lawn area and 150 m away a small wood grove. The radiative cooling process on a clear summer afternoon and evening, with sunset at 1915 hr, is shown in Table 5.1 (Landsberg, 1970).

The different cooling rates can be seen partly in Fig. 5.1. Particularly noteworthy is the fact that even 2 hr after sunset the wall and courtyard surfaces are 4–5°C warmer than the air temperature, and that is about 1°C higher in the courtyard than over the surrounding grass land. The grass surface itself, with its very low heat capacity cools most, already starting before sunset. A slight rise in the grass-surface temperature can be noted at 2020 hr. when dew formation started and the latent heat released briefly elevated the temperature. The process is essentially the story of the urban heat island in miniature on a clear, calm evening.

TABLE 5.1

Temperature Development in and Near Building Complex on Summer Afternoon and Evening[a]

Local standard time (hr)	Cld. oktas	Weather	Wind (msec⁻¹)	Air temp. (°C) (at 2-m ht.)			Surface temp. (°C)						
							Walls facing				Ground		
				C	G	W	N	E	S	W	C	G	W
1620	2	Sunshine	3	30.6	30.6	30	32	35	34.5	>44	>44	33	30
1934	3		1	28.3	27.8	28.2	30.5	31	31.5	33.5	33	29	27.5
2115	<1		<1	25.6	24.7	24.4	27.5	28	29.5	29.5	30.5	23	25

[a] C, courtyard; G, grass area; W, wood lot.

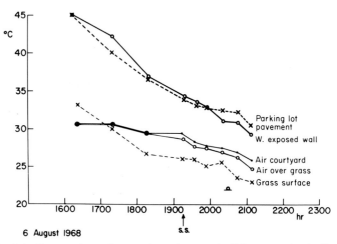

Fig. 5.1 Temperature changes in and near a building complex in the afternoon and evening hours. (S. S. marks time of sunset; small circle on line at 2030 hr marks time when visible dew was noted on grass.)

Similar observations were made around isolated shopping centers by Norwine (1972, 1973) and by Chopra and Pritchard (1972). Figure 5.2 shows a typical case of Norwine's observations on a calm, clear evening between 2200 and 2300 hr. A 2°C heat island is apparent. In contrast, on cloudy, windy days Norwine observed only a 0.5°C difference to the surroundings. In the cited cases there were no terrain conditions that could have produced such differences.

The maximum temperature difference when the heat island is pronounced shows up usually 2–3 hr after sunset. In the small-scale building complexes it generally disappears after midnight. Yet in large urban areas it is still notable at the time of sunrise and raises the minimum temperatures there. In midday, as we shall see later, the differences between the urban and rural areas are smallest.

The clearest evidence for the urban heat-island effect, in the absence of measurements prior to the establishment of a town, is the fact that the local temperatures in town have risen as it grew. This is, of course, predicated on the subtracting of general climatic trends in the region. There are a number of cases where this can be demonstrated. One of them is for Tokyo where the temperatures have been

Fig. 5.2 Mini-heat island in isolated shopping mall on a clear, calm March evening; isotherms in °F (from Norwine, 1972).

increasing since 1920 over and above the regional trend. This is shown in Fig. 5.3 according to Fukui (1970). He documented the changes shown in Table 5.2 for three rapidly growing towns and three smaller towns without notable urban growth. At the former, the change in a 30-yr interval was about 0.03°C yr^{-1}, at the latter it was only about 0.01°C yr^{-1}.

In the period after World War II when Tokyo was largely destroyed, but rapidly reconstructed, the urban temperature rose nearly 1°C. Daily maxima increased at a rate of 0.036 yr^{-1}, but the minima rose even faster by 0.047 yr^{-1} during the interval of reconstruction (1946–1963).

The same change of rapidly rising minima and more slowly rising maxima was placed on record for Paris by Dettwiller (1970a). This is graphically illustrated in Fig. 5.4. For a 78-yr interval (1891–1968), this author reported a rising trend of 0.011°C yr^{-1} for the daily maxima and 0.019°C yr^{-1} for the minima. The clear conclusion is that urbanization decreases the daily range of temperature. Paris

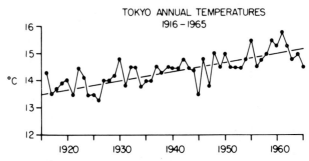

Fig. 5.3 Long-term annual temperature trends in Tokyo, Japan (after Fukui, 1970).

offers still more evidence for the rising urban temperatures. Philippe de la Hire (1640–1718) constructed a thermometer in 1670 and used as one of the calibration fix points the temperature in a cellar under the Paris Observatory (Middleton, 1966). Dettwiller (1978) attributed these first cellar observations to Jean-Dominique Cassini, installing a thermometer made by Mariotte in 1671. This cellar was 28 m below the surface and measurements have been made there since, first sporadically and later fairly regularly. Until about 1870 the temperature hovered between 11.7 and 11.9°C. Then a gradual rise was observed and in 1969 it was about 13.5°C. Thus the deep-soil temperature had risen about 1.7°C, a value very close to the 1.5°C rise in minimum temperatures shown in Fig. 32 in 78 yr. Dettwiller (1970a,b) also compared the central Paris mean temperatures during the 1951–1960 decade with surrounding stations at airports and in rural localities. The isotherms are shown in Fig. 5.5. The outlying weather stations had mean annual temperatures of 10.6–

TABLE 5.2

Temperature Rises in Japanese Cities 1936–1965[a]

Rapid growth		Slow (or no) growth	
City	Temp. rise (°C yr^{-1})	City	Temp. rise (°C yr^{-1})
Tokyo	0.032	Nemuro	0.005
Osaka	0.029	Tyoshi	0.011
Kyoto	0.032	Hikore	0.020

[a] After Fukui (1970).

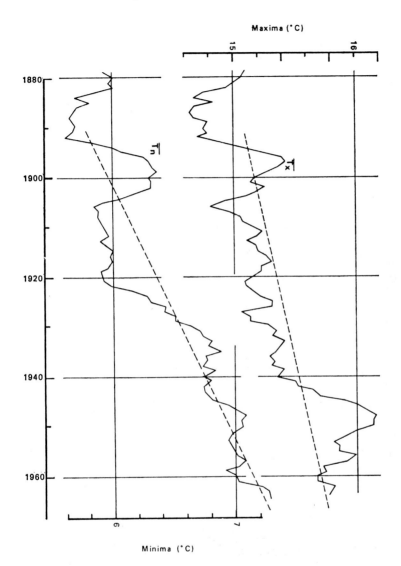

Fig. 5.4 Secular trends of daily maximum and minimum temperatures in Paris, France (from Dettwiller, 1970a).

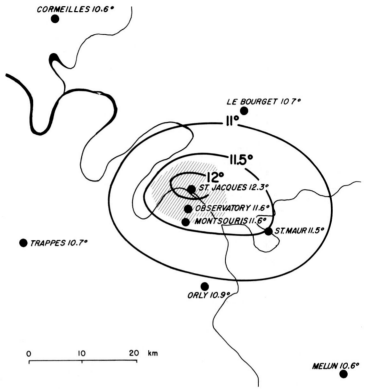

Fig. 5.5 Annual isothermes (°C) in the area of Paris, France (after Dettwiller, 1970b).

10.9°C, but in the center of Paris it was 12.3°C, about 1.5°C higher. These isotherms show how well justified the term "heat island" is. The values given by Dettwiller are all reduced to a uniform 50-m elevation to correct for possible orographic effects.

Annual temperature rises of similar magnitude have been found in other cities. Mitchell (1961a) made a comparison for a number of United States cities, using the most rapidly growing urban areas for the period of record when data were available for both the inner city and the rural environs. Part of his material is shown in Table 5.3. This Table shows values entirely compatible with those noted earlier for Paris and the Japanese cities. It is also interesting to observe that the winter values are considerably smaller than those found for summer. This is consistent with the view that the major cause for the

TABLE 5.3

Rate of Temperature Increase in Selected United States Cities as Excess over Rural Environs[a]

City	Period of record	Warming rate (°C yr⁻¹)			
		Winter	Spring	Summer	Autumn
Cleveland, OH	1895–1941	0.010	0.021	0.028	0.015
Boston, MA	1895–1933	0.016	0.018	0.026	0.021
Washington, DC	1893–1954	[b]	0.008	0.024	0.015
Tampa, FL	1895–1931	0.026	0.014	0.016	0.019
Baltimore, MD	1894–1954	[b]	0.011	0.017	0.015

[a] Adapted from Mitchell (1961a).
[b] value not significant; all other values significant at the 99 percent confidence level.

urban heat island is the change in the radiation balance and not rejected anthropogenic heat. Even the case of Tampa is not entirely incompatible with that view because there in winter are 12 percent more sunshine hours than in summer.

A comparison of the temperature in the city of Baltimore with that at the isolated rural seminary at Woodstock, Maryland, 25-km away, since the beginning of the century showed a growing temperature difference. The annual temperatures are now about 2°C higher than in the earlier period. The population of the metropolitan area grew from about ½ million to 2 million. As shown in Fig. 5.6, there is no sign of any trend at Woodstock, with the year-to-year fluctuations smoothed out by using 5-yr averages (consecutive, not overlapping). There is an indication of a levelling off of the urban–rural difference.

In a worldwide survey of urban data, covering the 90-yr interval 1871–1960, Dronia (1967) estimated an average urban temperature rise of 0.008 °C yr⁻¹. With the wide geographic dispersal of available data, there is a satisfactory agreement of the magnitude of urban influence on air temperatures. One can also persuasively demonstrate that as urbanization progresses, the outskirts or suburban areas become gradually incorporated into the heat island. A good case has been made for Kew Observatory near London (Moffitt, 1972). Although the observatory is located in a parklike area it is affected by the spreading of solidly built-up neighborhoods beyond the park's confines. The temperatures there were compared with those of

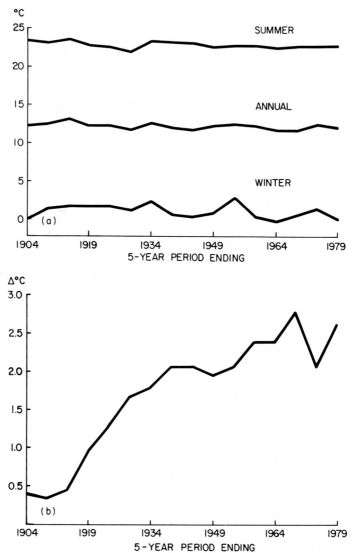

Fig. 5.6 Temperature trends (5-yr consecutive means) at rural reference station Woodstock, Maryland (upper panel), and evolution of temperature difference Baltimore, Maryland, downtown area and Woodstock.

Rothamsted Agricultural Experiment Station, completely unaffected by urbanization. It was found that the temperature in 1880 was about $+0.74°C$, almost to a tenth of a degree of what would be expected by elevation difference. In the 1960s the difference had risen by 1°C or about 0.013°C yr^{-1}.

Most attention in the literature has been devoted to the nocturnal heat island and, more specifically, to the maximum difference that would develop between the city and the environs. There are examples from urban areas of various sizes, from all continents, and with a wide variety of topography. It is now general practice to show isolines of the temperature differences in the city to a location outside that is assumed to be uninfluenced by the urbanization. A few typical examples will suffice to give the essential features of such observations.

The evolution of urban heat island is rapid. In Columbia, Maryland, in 1968, when the town had about 1000 inhabitants, the maximum difference in a residential area was 1°C. Only the beginning of a business center with a large parking lot and several office buildings showed $\Delta T = 3°C$ (Fig. 5.7). By 1974, with a little over 20,000 inhabitants, the maximum heat island had grown to $\Delta T = 7°C$ (Fig. 5.8). That value was almost exactly the same as observed by Hutcheon *et al.* (1967) in Corvallis, Oregon, a town of 21,000, i.e., 6.7°C (Fig. 5.9). Before commenting on these population-related values, one more example of an urban heat island has to be presented. This one reflects conditions in the London metropolis, Fig. 5.10 (Chandler, 1965). In contrast to the previous examples it depicts minimum temperatures and thus does not show the greatest extent of the heat island. The isotherms also give observed values rather than temperature differences. These isotherms are probably far too smooth to represent reality. Observations in other large cities suggest a far more detailed structure. However, the figure reflects one feature common in well-developed heat islands of large cities, and that is the crowding of isotherms at the edge of the built-up area. This steep gradient of nocturnal temperatures at the urban border has some of the characteristics of a meteorological cold front. Often the greatest differences are observed in the long winter nights when the sky is clear and the air calm. Good examples have been cited by Schulze (1969), and a very detailed analysis for the evolution of such a case has been given for Berlin by Reichenbächer (1978).

There have been a number of attempts to relate the magnitude of

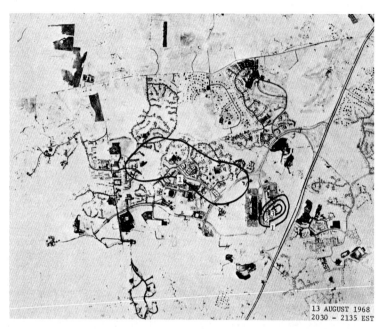

13 AUGUST 1968
2030 – 2135 EST

Fig. 5.7 Maximum heat island at Columbia, Maryland, in 1968 at the beginning of urban construction. (From Landsberg, 1975.)

the heat island to city size. To find a convenient parameter to reflect the latter has not been easy. As early as 1953 Mitchell noted that a substantial amount of the variance of temperature rise in cities could be explained by a function of population growth. The fastest-growing cities showed the highest warming rates. He chose the square root of the population number as the most representative factor for the urban contribution to the temperature change (Mitchell, 1961). In his sample of 77 United States cities he could show a trend ($r = 0.86$) but less to winter data ($r = 0.59$).

Population number is not a physical quantity, but other parameters such as area, degree of alteration of surface conditions, and heat production cannot be assessed entirely unambiguously. Hence one has to settle for dimensionly deficient models. Oke (1973) pursued this matter experimentally by relating measurements of the urban–rural temperature difference (ΔT_{u-r}) to population, starting with nocturnal field trips in the Montreal area. He used 10 settlements

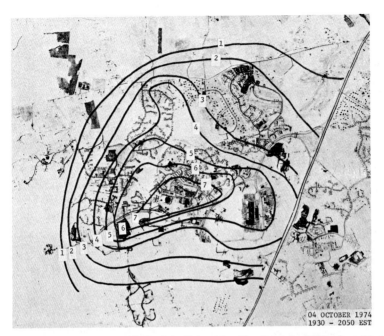

Fig. 5.8 Maximum heat island at Columbia, Maryland, in 1974 when town had grown to 20,000 inhabitants. (From Landsberg, 1975.)

of sizes between 1000 and 1 million inhabitants in a region, where topographic elements were of minimal influence. The measurements were made about 3 hr after sunset on days with clear skies and light winds. This showed the conditions of maximal development of the heat islands. He found that he could represent his data by an empirical relation

$$\Delta T_{u-r} = \frac{0.25 P^{1/4}}{\bar{u}^{1/2}} \tag{5.1}$$

where P is the population figure and $\bar{u}$ is the mean wind speed.

This relation explained over 70 percent of the variance with a standard error estimate of $\pm 1.6°C$. The awkwardness of the relation becomes immediately obvious when one considers calm conditions. That means $\bar{u} = 0$ and mathematically the temperature difference would become indefinite. As it turns out, under those conditions (Oke, 1976), the $\Delta T_{(u-r)max}$ is proportional to the logarithm of the

Fig. 5.9 Heat island at Corvallis, Oregon, at about 2200 hr on a January night (adapted from Hutcheon *et al.*, 1967). Isotherms in °C are superimposed upon an aerial photograph (picture credit Western Ways, Inc., courtesy of W. P. Lowry).

population. Oke was able to use this relation to represent the maximal urban heat islands very well. Interestingly enough there were differences between North American and European settlements. For his ten settlements in Quebec and eight other towns and cities Oke determined a regression equation

$$\Delta T_{(u-r)max} = 3.06 \log P - 6.79 \qquad (5.2)$$

This explained 96 percent of the variance with a standard error of ± 0.7°C.

The relation was quite independently confirmed by the data (discussed above and represented in Fig. 5.8) collected in the growing community of Columbia, Maryland (Landsberg, 1975). The maximum heat-island values measured there between 1967 and 1974, and the corresponding population figures, fitted Oke's regression exactly

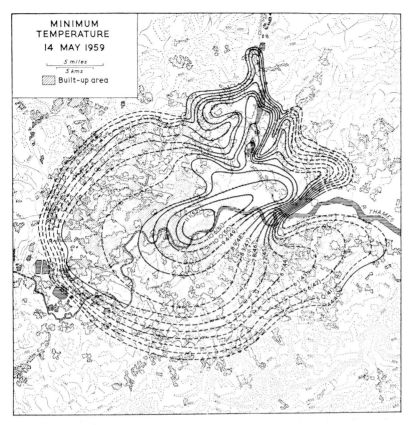

Fig. 5.10 The heat island of London shown by isotherms of minimum temperatures (both °C and °F). Dashed lines are uncertain (from Chandler, 1965).

(Fig. 5.11). His regression for 11 European cities was

$$\Delta T_{(u-r)max} = 2.01\ P - 4.06 \qquad (5.3)$$

The variance explained by this relation was considerably lower than in North America, namely 74 percent, and the standard error was ±0.9°C. The increment of the heat island with the population figure is smaller in Europe than North America.

Measurements of heat islands in the Tropics and in the Arctic seem to follow their own rules. In the Arctic intense inversions forming in slight depressions make comparisons difficult, but a first

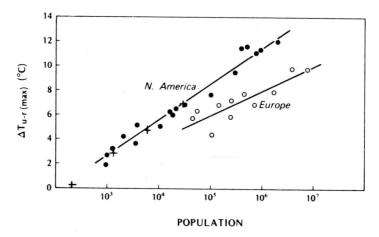

POPULATION

Fig. 5.11 Relation of the maximum urban heat island (ΔT_{u-r} °C) as related to population, both for European (open circles) and American cities (solid circles). The two crosses refer to the Columbia, Maryland, heat island shown in Figs. 5.7 and 5.8 (from Oke, 1979).

impression is that there heat islands are more intense than population size would lead one to postulate.[2] In contrast, in the tropical cities from which data are available, the maximum ΔT_{u-r} values are less than population size would suggest. In many of those cases topographical complications defy such simple rules.

Clearly, the extent and intensity of the urban heat island is also profoundly influenced by topography. Most investigators have tried to eliminate elevation differences by making a correction with an assumed average lapse rate, but this can not compensate for air drainage. As we shall see later, secondary circulations affect the heat island in complex topography. Even without that, there are no-

---

[2] A report from the University of Alaska (1978) states under the heading *Heat Islands* the following: "The heat island associated with the city of Fairbanks, Alaska, was studied as a means of isolating the effects of self-heating and modified radiation transfer. The observed steady-state temperature difference was found to be about 10°C, with transient values reaching 14°C (25°F). This high value is probably due to the extremely steep ground inversions known to exist in Fairbanks, as the heat island intensity correlated well with how much warmer it was at the 60-meter elevation than at ground level over the city. The depth of the mixing layer was less than 90 meters, but the temperature structure at higher levels was disturbed, apparently by coherent lifting of the stable air."

table influences of large rivers bisecting urban areas. For example, the Potomac River in Washington, D. C., causes in summer a split in the daytime heat island with a core in the business section of the District of Columbia and other cores on the Virginia side of the river in Arlington and Alexandria. The broad belt of green surfaces on either side of the river contributes to that condition (Landsberg, 1950). The same phenomenon was discovered by Emonds (1954) for Bonn and Beuel on opposite banks of the Rhine river.

It might be well to add here a further cautionary note on the role of interaction between the heat-island effect and urban topography. This refers to a classical micrometeorological nocturnal winter traverse made by Middleton and Millar (1936) in Toronto. This set of measurements was made on a calm, clear February night, shortly after midnight (Fig. 5.12). The trip was along Yonge Street northward from the shore of Lake Ontario to the Don River and back. At the lakefront there were relatively high temperatures of − 9.4°C. In the densely built-up section there was a rise about 1-km inland of about 0.7°C, evidently a heat-island effect. Further inland, with the terrain rising about 100 m, there was a gradual temperature drop, with an estimate of the rural, or at least suburban, temperature of − 12°C. But then orographic cuts appeared with extraordinary inversions and a low temperature of − 26°C. In such a case it is very difficult to decide what is lake effect, what is terrain effect, and what is attributable to urbanization. Another disturbance of anthropogenic origin was noted by Fukuoka (1980) in Tokyo. There the soil temper-

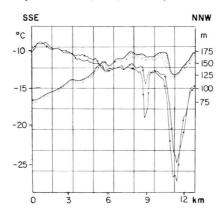

Fig. 5.12 Temperature cross section on a winter night at Toronto as related to topography (adapted from Middleton and Millar, 1936).

atures at 60- and 90-cm depth above a subway in compacted material were about 6°C higher than 30-m distant from the subway tunnels under shrubbery.

5.2 FLUCTUATIONS OF THE HEAT ISLAND

The heat island is not a constant condition. It shows both periodic and aperiodic fluctuations. The diurnal variation is very pronounced. In daytime the urban–rural difference even on sunny, calm days is generally quite small. In fact, in some instances measurements at an airport near the city show higher values than parts of the urban area. The typical diurnal temperature variations on a clear day are reflected in the thermographic records of the city and airport station at Richmond, Virginia. Essentially, the differences are negligible in daytime but develop rapidly after sunset and stay throughout the night (Fig. 5.13).

It is instructive to look at the frequency distribution of differences of the daily extreme temperatures at different times of the year. An example for the year 1953 at Lincoln, Nebraska, is shown in Fig. 5.14. At that time the city had about 100,000 inhabitants. In the warm season the daily maxima at the airport had a tendency to be higher than those recorded in the city. This reflected the microcli-

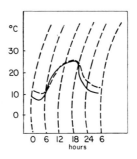

Fig. 5.13 Diurnal variation of temperature on calm, clear day in Richmond, Virginia. Solid line represents rural conditions at airport, dashed line gives urban conditions.

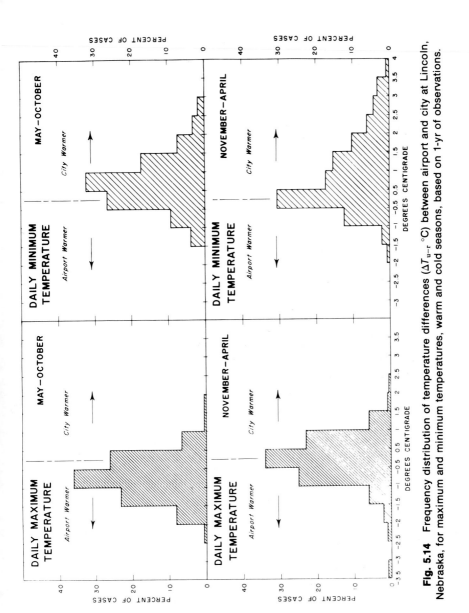

Fig. 5.14 Frequency distribution of temperature differences (ΔT_{u-r} °C) between airport and city at Lincoln, Nebraska, for maximum and minimum temperatures, warm and cold seasons, based on 1-yr of observations.

mate of the station exposures. In the city the parklike setting of the observing place on the University campus stayed cooler than the expanses of runways at the airport. However, at night the city was, in the overwhelming number of cases, warmer than the airport, as shown by the minimum temperatures.

In the cold season conditions are quite different. The differences in maximum temperatures are quite evenly distributed around zero departure, which is the most frequent case. The dormant trees at the campus location have lost their influence and the symmetrical shape of the frequency diagram suggests that the temperature differences of city–airport are random variations. The nocturnal differences are quite dissimilar. Zero differences are still the most frequent category, but the number of times when the city had higher minima than the airport is vastly preponderant over times when they were lower in town.

From Chandler's (1965) studies in London one can see a corresponding picture for that metropolis, as shown in Fig. 5.15. The data represented there cover a decade of observations at the city site in Kensington and the rural station at Wisley. Figure 5.15 gives the mean annual number of days with differences in maximum and minimum temperatures of various magnitudes. Again, the maxima yield a far more symmetrical picture than the minima, although the modal value is $+0.6°C$. Thus the heat island is still measurable on many occasions during the daytime in an urban area of London's size, but again the real difference shows up in the minima, which show overwhelmingly positive departures. Unwin (1980) has analyzed the Birmingham, England, heat island according to synoptic flow patterns, using 11 weather types. The anticyclonic type has the greatest preponderance of heat-island development, the least occur in cyclonic weather. Clarke and Peterson (1972) have presented maps for the whole United States, showing medians, quartiles, and deciles of heat islands based on model calculations.

A very instructive example of the development of the heat island has been placed on record by Oke and East (1971). They show the hourly warming and cooling rates on a winter day in Montreal (as given in Fig. 5.16) for urban, suburban, and rural sites. There were very few or no clouds and very low wind speeds in the night hours. Warming evidently ceased after 1600 hr in the suburban and rural areas, with the latter experiencing a rather abrupt increase in cooling rate even 1 hr before sunset. The urban area did not start cooling

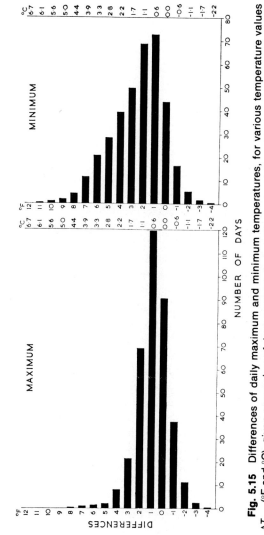

Fig. 5.15 Differences of daily maximum and minimum temperatures, for various temperature values ΔT_{u-r} (°F and °C), given as number of days per year, from a 10-yr record (from Chandler, 1965), for London, England.

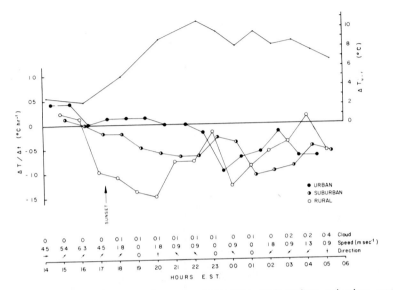

Fig. 5.16 Warming or cooling rates $\Delta T/\Delta t$ (°C hr^{-1}) for urban, suburban, and rural areas in and near Montreal, Quebec, Canada, during a February night and the simultaneous heat island intensity T_{u-r} (°C) (from Oke and East, 1971).

until 2200 hr. After midnight the cooling rates in all three environments had become nearly equal and hovered irregularly around 0.5°C hr^{-1}. The development of the heat island is shown in the upper part of the diagram. It reached its maximum value $\Delta T_{u-r} = 10.5$°C at 2200 hr and gradually dropped to $\Delta T_{u-r} = 6.5$°C by 0500 hr.

There is indeed a marked difference in the diurnal variation of ΔT_{u-r} values between summer and winter. Hage (1972) has shown this for the nights with intense heat islands in Edmonton, Alberta. Data from two airports there were available, one in town, 3 km from the center, the other 26 km to the southwest. Comparisons in two winters, smoothed by a 5-hr running mean, showed maximum heat-island development around 2100 hr. The difference ΔT_{u-r} decreases till 1300 hr on the following days (Fig. 5.17a) but it stays positive. Hage suggests that this is due to the anthropogenic heat release in the city. In the warm season (Fig. 5.17b) the daytime values are very close to even until about 1700 hr when the cooling starts rurally, but ΔT_{u-r} does not reach its maximum value until midnight.

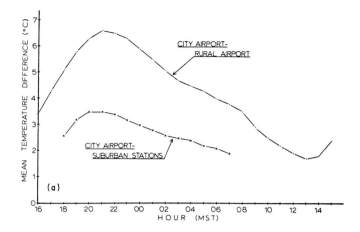

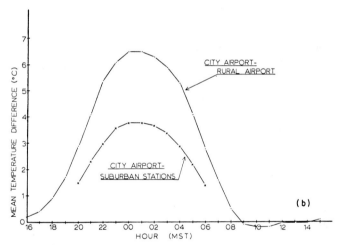

Fig. 5.17 Five-hourly running means of temperatures for nights with well-developed heat island at Edmonton, Alberta, Canada, for 2 yr. Upper panel (a) for December, January, February; lower panel (b) for May, June, July (from Hage, 1972).

It is noteworthy that in winter and summer the averaged maximal values of ΔT_{u-r} reach about the same value of about 6.5°C.

There is also a report by Nkedirim and Truch (1981) from Calgary, Alberta, that is based on 1 yr of observations and that shows a very unique diurnal variation. Using a mean value of ΔT_{u-r} these authors show maximum mean values at about 9 a.m. ($\Delta T_{u-r} = 8$°C) and minimum mean values at 1600 hr ($\Delta T_{u-r} = 5.5$°C). They attribute that to the peak in fuel use that occurs in Calgary in the early forenoon. There is no other case of this type mentioned in the literature. Nkedirim and Truch also give a table of mean monthly intensities of ΔT_{u-r}, which we reproduce here (Table 5.4). The values look somewhat ragged, presumably because in a 1-yr series the incidence of clear or cloudy, calm or windy, days is not sufficiently smooth from month to month as in a longer record. Nonetheless, the cold season has substantially higher ΔT_{u-r} values than the warm season. From the ratio of the anthropogenic heat release to observed solar radiation one can infer that a part of the heat island is anthropogenic. In summer, however, that element is negligible and one must attribute all of it to absorbed solar radiation. The fact that the summer values are lower is probably attributable to local wind currents, which develop in the complex topography near the town.

On the issue of solar versus anthropogenic origin of the heat island, another periodic fluctuation needs consideration. In a comparison of daily extreme temperatures between city and airport in New Haven, Connecticut, Mitchell (1961b) found a weekly variation. Data

TABLE 5.4

Annual Variation of Mean Monthly Heat-Island Intensity for 1974–1975[a,b]

Month	ΔT_{u-r} (°C)	Ratio (%)[c]	Month	ΔT_{u-r} (°C)	Ratio (%)[c]
I	10.1	43	VII	5.8	0.1
II	7.8	19	VIII	5.2	1
III	5.9	10	IX	4.6	3
IV	9.4	4	X	4.3	9
V	6.8	2	XI	8.7	24
VI	7.8	1	XII	10.1	55

[a] After Nkedirim and Truch (1980).
[b] For Calgary.
[c] Ratio of artificial heat release to observed solar radiation.

TABLE 5.5

Temperature Difference between City and Airport at New Haven, Connecticut, in the Winters (1939–1943)[a]

Day	ΔT_{u-r} (°C)		
	max	min	Daily mean
Monday	0	1.1	0.6
Tuesday	0.1	1.3	0.7
Wednesday	0	1.2	0.6
Thursday	0	1.3	0.7
Friday	0	1.3	0.6
Saturday	0	1.2	0.6
Sunday	0.1	0.6	0.3

[a] After Mitchell (1961b).

for four winter seasons yielded the values for ΔT_{u-r} given in Table 5.5. We note again that there is basically no difference for the (daytime) maxima. In the minima there appears some weekly variation. The Sunday value is only about half of the difference values on other days. Mitchell suggested that this result (significant at the 99 percent level) is caused by the fact that the city is "dormant" on Sundays. Landsberg and Brush (1980) repeated the analysis of the weekly fluctuations of the heat island in Baltimore, Maryland, by a comparison of the urban temperatures with those of the Baltimore–Washington International Airport (BWI) during the 5-yr interval 1971–1975. This survey comprised both winter and summer months. The results are shown in Table 5.6. In winter the weekdays were 0.8°C warmer in the city than at the airport, but on Sundays there was only a 0.2°C difference. Statistically the difference between Sundays and the weekdays is significant at the 99 percent level. In summer the Sundays were warmer than the weekdays, but there was no significance between the Sunday and Monday values. For an as yet unexplained reason all other weekdays had lower values. However, these data show again that the average summer heat island value is about 1°C higher than the winter values. This is the result of higher winter wind speeds and higher net radiation values in summer.

TABLE 5.6

Daily Temperature Differences (°C) by Days of the Week for Baltimore City and Baltimore–Washington International Airport (1971–1975)

Day	Winter	Summer
Sunday	0.2	2.2
Monday	0.9	2.0
Tuesday	1.0	1.3
Wednesday	0.8	1.6
Thursday	0.7	1.4
Friday	0.7	1.4
Saturday	0.4	1.7
Mean	0.7	1.6
σ	1.7	1.7

5.3 VERTICAL EXTENT OF THE HEAT ISLAND

The notable surface temperature anomalies of the urban space must of necessity show an effect in the vertical dimension. This is again most notable at night. It was clearly demonstrated in a classical study by Duckworth and Sandberg (1954) in the San Francisco Bay area. They made 32 parallel ascents with a captured balloon in urbanized and nearby rural sectors. In 30 of these the rural locality showed a radiatively formed ground inversion. At the urban sites there was a nearly isothermal condition in lowest 100 m. A typical case is shown in Fig. 5.18. This also shows the not uncommon phenomenon that above the shallow rural ground inversion it is warmer there compared with the values given by the urban sounding. This has been called the *crossover effect*.

Although ascents in and near urban areas show similar vertical temperature structures at night, the strong stability over rural areas and the relative instability over the city or somewhat downwind from the city is more pronounced in simple flat terrain. Where topographic complexity prevails topoclimatic differences appear with cold-air drainage to low spots in the landscape. In slightly rolling

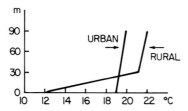

Fig. 5.18 Typical nocturnal vertical temperature structure over an urban and adjacent rural area, showing the so-called crossover effect.

terrain the heat-island effect may persist, but the intricacy of vertical temperature structure under such conditions can be seen in Fig. 5.19 for Cincinnati, Ohio (Clarke and McElroy, 1970). It is therefore quite unreasonable to expect uniform or typical conditions in all urban areas (Tyson *et al.*, 1972). Yet the crossover effect is suffi- ciently common to deserve further attention. A longer series of ob- servations, rather than a few samples from ascents, is available from tower observations. In Vienna, Austria, comparisons of tempera- ture conditions were made at five heights up to 110 m on the cathe- dral tower in the urban center and a tower in an unsettled area 5-km distant (Machalek, 1977). The results of this study did show the crossover effect on frequent occasions. It also pointed up the no- table influence of the "second urban surface," namely the average roof level, which showed in autumn conditions frequent inversions between 25 and 50 m. This level will be an important interface in densely built-up areas with relatively uniform building height.

Several factors combine to cause the nocturnal upper inversion over the city and the fact that it is at times colder at the same eleva- tion than over the rural area. There the intense surface inversion will cause some subsidence and consequently some adiabatic warming in the layer just above the inversion. In the city there is the formation of a pollutant layer, somewhat lifted by the low-level instability. This layer can cause heat loss by outgoing radiation and hence cool below the temperature prevalent at the same height in the cleaner rural air column (Böhlen, 1978). Model experiments, however, suggest (Atwater, 1975) that the temperature changes caused by ra- diatively active pollutants remain a relatively minor factor in the development of vertical temperature structure.

In daytime under sunny conditions the lapse rates in urban areas rapidly steepen, especially in the warm season. Helicopter ascents in the Cincinnati area showed in summer, even before noon, super-

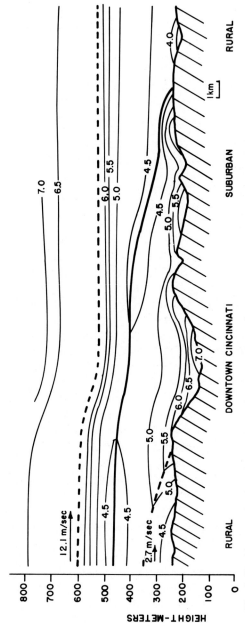

Fig. 5.19 Nocturnal temperature structure over Cincinnati, Ohio, in rolling terrain with lifted-off inversion over city (after Clarke and McElroy, 1970).

adiabatic gradients in the lowest 100 m and nearly adiabatic conditions in the next 300 m (McCormick and Kurfis, 1966). In winter the steep lapse rates develop more slowly, and isothermal or inversion conditions aloft at relatively shallow heights may persist throughout the day (Klysik and Tarajowska, 1977).

Ludwig (1970) established a rather close relation between the magnitude of the nocturnal urban–rural temperature differences and the vertical lapse rates. These lapse rates were obtained from the aerological soundings at locations in or near the cities. In some instances these were at considerable distances from the city, hence there is considerable scatter in the data. Ludwig split his data according to urban size into three groups and established regressions between the lapse rate and the magnitude of the heat island. The regression lines are shown in Fig. 5.20. The corresponding regression

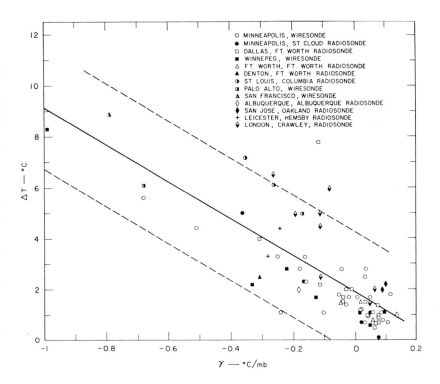

Fig. 5.20 Temperature difference ΔT_{u-r} (°C) as function of low-level lapse rate (from Ludwig, 1970).

equations are:

$$\Delta T = 1.3 - 6.78\Gamma \qquad (r = -0.95) \qquad [<\tfrac{1}{2} \times 10^6 \text{ inhabitants}]$$
$$(5.4a)$$

$$\Delta T = 1.7 - 7.24\Gamma \qquad (r = -0.81) \qquad [\tfrac{1}{2} \text{ to } 2 \times 10^6 \text{ inhabitants}]$$
$$(5.4b)$$

$$\Delta T = 2.6 - 14.8\Gamma \qquad (r = -0.87) \qquad [>2 \times 10^6 \text{ inhabitants}]$$
$$(5.4c)$$

where ΔT is the magnitude of the urban–rural temperature difference (°C) and Γ is the the vertical lapse rate (°C/mbar) (negative values indicate stability or inversion conditions, 0 value is neutral stability or adiabatic lapse rate, and positive values indicate vertical instability).

As Ludwig points out both the vertical lapse rates and the nocturnal temperature differences are manifestations of the same processes, principally the different radiation fluxes in the urban and rural areas.

5.4 WIND INFLUENCE ON THE HEAT ISLAND

Of all the synoptic parameters the wind has the greatest influence on the development of the nocturnal heat island. Although clear sky is an important contributing factor, experience has shown that a strong wind will supercede the radiative fluxes. A first attempt to quantify this effect was made by Summers (1964). His results in Montreal led to the following relation between the wind and the strength of the heat island ΔT_{u-r}

$$\Delta T_{u-r} = \left(2r \frac{\partial \Theta}{\partial z} Q_u\right) \Big/ (\rho c_p u) \qquad (5.5)$$

where

r distance from upwind edge of the city to the center
$\partial\Theta/\partial z$ potential temperature Θ increase with height z
Q_u urban excessive heat per unit area
ρ air density

c_p specific heat at constant pressure
u wind speed

It must be pointed out that the vertical increase of potential temperature with height has its own urban component and can be zero, which will make the formula inapplicable. If upwind values are used the rural inversions will inflate the value.

Oke (1976) tested the relation in Vancouver and found that the relation explained only 27 percent of the variance. My experience indicates that each city, because of topographic differences, reacts differently to the wind. In many instances the heat-island development is different for various wind directions.

Oke's (1976) graphical representations for Vancouver and Quebec observations are instructive (Fig. 5.21), and the scatter of the Vancouver data, which depends on the Summers parameter, gives a measure of the uncertainty (Fig. 5.22). Oke and East (1971) showed the vertical distribution of the Montreal urban heat island for a few observations in the lowest kilometer for weak and somewhat stronger winds (Fig. 5.23) and the difference is quite startling. Although a vestige of the heat island exists even under the stronger winds, it de-

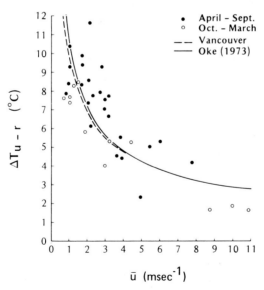

Fig. 5.21 Heat-island intensity ΔT_{u-r} (°C) 1–3 hr after sunset in Vancouver (solid line and points), dashed line based on Quebec data (from Oke, 1976).

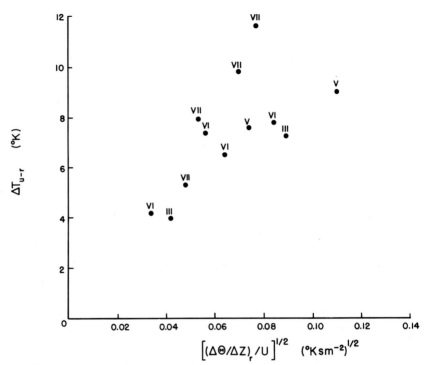

Fig. 5.22 Heat-island measurements in Vancouver (1972–1973) in relation to the Summers model parameter (abscissa). Roman numerals indicate month (after Oke, 1976).

creases very gradually and the standard deviation of the observations decreases with height. In contrast, the weak winds show a very rapid decrease to the 200-m level and the upper inversion at 600 m with increasing variability of the observations above that level.

A systematic analysis of the data from the extensive network of temperature observations in the Washington, D. C., metropolitan area by Nicholas (1971) showed the wind influence clearly. In a strong northwest wind (10 m sec^{-1}), after passage of a cold front, the minimum temperatures of the metropolitan area became very uniform. There was no sign of the influence of the river, the 100-m differences in the topography, or even the anthropogenic heat production in the winter night. There was no effect of the clear sky during that night (Fig. 5.24a). In contrast, during the same season, a night

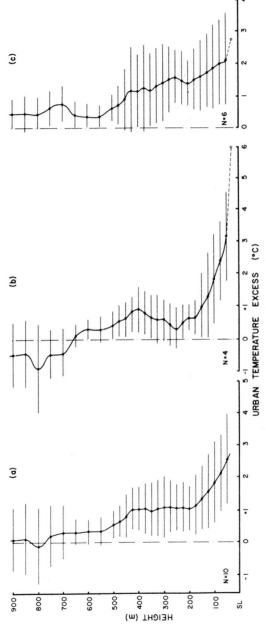

Fig. 5.23 Height variation of urban temperature excess (°C) just after sunrise for (a) all wind speeds, (b) 0–3 m sec⁻¹, (c) 3–6 m sec⁻¹. Bars show 1 standard deviation (from Oke and East, 1971).

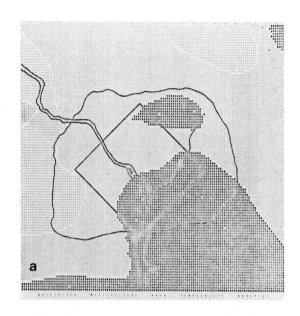

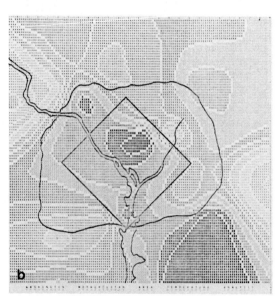

Fig. 5.24 Symaps of minimum temperature distribution in Washington, D. C., area (a) on a night with strong winds and (b) one with weak winds (from Nicholas, 1971). Note lack of contrast in (a) and crowding of isotherms in (b) (from Nicholas, 1971).

with winds <3 m sec^{-1} caused a 6°C heat island, also under a clear sky (Fig. 5.24b).

The observations quite clearly show a threshold of wind speed that prohibits the heat-island development. This varies from urban area to urban area. Oke and Hannell (1970) related this to city size, using population data from a few cities as their measure (from 3.3×10^3 to 8.5×10^6 people). Their material, shown in Table 5.7, shows that in small towns winds as light as 4-m sec^{-1} can eliminate the heat-island effect. Oke and Hannell (1970) place the limiting wind speed at

$$u_{lim} = 3.4 \log P - 11.6 \qquad (5.6)$$

where P is the population number. The variance explained by this empirical relation is a remarkable 94 percent.

These observations have significance for local weather forecasters. They have in recent years taken note of the urban–rural temperature differences. The predictions will either give a range for a metropolitan area or else give operative values for the inner city and the suburbs. But often the forecasts give far too large a range by ignoring the wind-speed factor. Most of these errors in the eastern North American cities occur in winter after passage of sharp cold fronts, which are usually followed by brisk winds. In the first postfrontal night, suburban temperatures are often predicted too low. Only after the cold air mass settles on the second or third postfrontal night and winds calm down does a notable heat island develop.

TABLE 5.7

Limiting Wind Speeds for Occurrence of Heat Island in Cities of Different Size[a]

City	Population ($\times 10^3$) at time of survey	Limiting wind speed (m sec^{-1})
London, England	8500	12
Montreal, Canada	2000	11
Bremen, Germany	400	8
Hamilton, Canada	300	6–8
Reading, England	120	4–7
Kimagaya, Japan	50	5
Palo Alto, United States	33	3–5

[a] After Oke and Hannell (1970).

5.5 CONSEQUENCES OF THE URBAN HEAT ISLAND

The urban heat island has a number of meteorological and several socioeconomic impacts, some beneficial, others detrimental. In cold climates and in many areas in winter the heat island will obviously reduce the need for heating. One can see this to a first approximation by comparison of annual heating degree days[3] observed at urban and neighboring airport stations. Although this will depend on the relative location and distance of each observing station, the values given in Table 5.8 represent fairly realistically the fuel needs in the urban and rural environments of the respective localities.

If one excludes the unreasonably large value for Los Angeles, the average difference is about 8 percent. A similar compilation made in the 1960s based on the then used 30-yr comparison interval 1931–1960 yielded about 9 percent for 18 pairs of stations. Again excluding Los Angeles, the average difference is about 220 degree days. In 1956 Kratzer cited 160 degree-day differences in various sectors of Berlin. The approximate gradient in United States cities can be seen in Table 5.8 for two localities. In New York the urban location is Central Park, the airport is John F. Kennedy International Airport; but the value for La Guardia Airport, which is still within the heat island, is also given in parentheses. For Chicago the comparison is between the University station and O'Hare Airport with the value for Midway Airport, also under urban influence, shown in parentheses.

The reduction in heating degree days will, of course, result in fuel economy, depending on the volume of the dwellings to be heated. The closer houses are built together the more protection against the weather demands they offer each other and the less is the relative fuel consumption. Table 5.9 illustrates this for single-family housing.

Even though the heat island contributes to fuel savings in the cold season, in many places this is outweighed by the additional energy needs for air-conditioning in summer. Using the same station pairs

---

[3] Degree days are based on a heating threshold of 18°C (65°F). All daily mean temperature values below this value are subtracted from it and the differences summed for the heating season.

TABLE 5.8

Heating Degree-Day Differences[a] between Urban and Airport Stations (1941–1970 average)

	Heating degree day			
City	Urban	Airport	Difference	%
Los Angeles, CA	692	1011	319	46
Charleston, SC	1078	1192	114	6
Baltimore, MD	2278	2627	349	15
Washington, DC[b]	2339	2467	128	6
St. Louis, MO	2492	2639	147	6
Seattle, WA	2493	2881	388	16
Kansas City, MO	2641	2867	226	9
New York, NY	2693	(2727)[c] 2880	187	7
Denver, CO	3058	3342	284	9
Chicago, IL	3371	(3403)[d] 3609	283	7
Detroit, MI	3460	3556	96	3
Albany, NY	3552	3827	275	8

[a] Base 18°C.
[b] In Washington, D.C., the National Airport is close to the middle of the metropolitan heat island. It was therefore chosen as the "urban" station. For comparison, the station College Park in suburban Maryland at the edge of the urban heat island is given in the "Airport" column.
[c] La Guardia Airport.
[d] Midway Airport.

TABLE 5.9

Relative Fuel Consumption of Single-Family Dwellings in Various Settlement Densities

Type of housing	Relative fuel needs
Dense urban row housing	1
Town houses, open rows	1.3
Duplex housing, free on all sides	1.5
Single home	
free on all sides, with adjacent houses close by	1.7
free standing	2.1

TABLE 5.10

Cooling Degree-Day Differences[a] between Urban and Airport Stations (1941–1970 Average)

City	Cooling degree days		Difference	%
	Urban	Airport		
Seattle, WA	111	72	39	35
Albany, NY	366	321	45	12
Detroit, MI	416	366	50	12
Denver, CO	416	350	66	16
Chicago, IL	463 (518)[b]	372	91	20
New York, NY	598 (586)[c]	482	116	11
Los Angeles, CA	663	344	319	48
Washington, DC	792	650	142	18
Baltimore, MD	835	620	215	14
Kansas City, MO	857	795	62	7
St. Louis, MO	918	826	92	6
Charleston, SC	1318	1164	154	7

[a] Base 18°C.
[b] Midway Airport.
[c] La Guardia Airport.

as for heating degree days, Table 5.10 lists the cooling degree days, this time counting the sum of the daily excess temperatures above the 18°C base.

Again eliminating Los Angeles, and also Seattle, from consideration because their airports are under ocean influence in summer, the average urban increase in heat load is about 12 percent. Because energy requirements for air-conditioning are higher than for heating, this accrues to a distinct disadvantage for the urban dweller. Moreover, air-conditioning aggravates the heat island because the equipment discharges the heat to the outside air where it mixes with air already warmed by the hot air forming adjacent to sunlit walls and pavements.

In Table 5.10 also, a few cases deserve comment. La Guardia Airport in New York City hardly differs from the Central Park location and one can safely assume that conditions in the tenement districts are still worse. In Chicago the University station shows a lower value than the Midway Airport. This is caused by the lake breeze, which in summer frequently cools the University area but does not penetrate as far inland as the Midway Airport.

The effect of the urban heat island on snow frequency and snow covers is very pronounced. For example, a survey in the area of New York City showed that in lower Manhattan snow probability on winter days favoring precipitation is less than 30 percent, compared with 40–45 percent in outlying sectors of the metropolis (Grillo and Spar, 1971). A particularly instructive example has been placed on record by Lindquist (1968) from Lund in Sweden. In this town a moderate snowfall was observed in the winter of 1965. In the countryside between 6 and 8 cm of snow were measured but in the town center only 3 cm fell. The temperature during midday was about 0.5°C in the town center and −1.0°C in the countryside. The isolines of equal snowfall are almost concentric in and around this compact town. This case is shown in Chapter 8 (see Fig. 8.12).

The heat-island effect is also very pronounced in the average dates of last freezing temperature in spring and the first freezing temperature in autumn in urban areas. In the city center the former may occur several weeks earlier and the latter several weeks later than in suburbia or the countryside. For Washington, D. C., the average date of the last freezing temperatures in the central city is about 3 weeks earlier than in the outlying districts (Fig. 5.25). Phenological observations show that the magnolias there bloom about two weeks earlier than in the suburbs. In autumn the city has, on an average, the first freezing temperature about November 3. In the outlying suburbs 0°C will usually be observed two weeks earlier. Thus the freeze-free season is about 35 days longer than in the rural sector. Davitaya (1958) noted for Moscow, U. S. S. R., an increase of 30 days in freeze-free days during the year, and Kratzer (1956) reported for Munich, West Germany, a 61-day extension of the freeze-free season over the rural environs.

Freezing temperatures have a number of implications for urban areas. They affect snow removal, and passages through the freezing point are important for frost heaving, pot holes, and damage to structures. Hershfield (1971) has made a comparison of the frequency of freeze–thaw cycles at Baltimore City and at the Baltimore Airport, 12-km distant. He used two ranges, based on daily maximum and minimum temperature observations, namely 31–32°F (−0.5–0°C) and 30–34°F (−1 to +1°C). The annual frequencies for a 10-yr

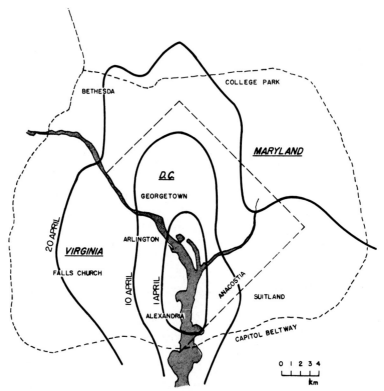

Fig. 5.25 Average date of last freezing temperature in spring in Washington, D. C.

period are shown in Fig. 5.26. The freeze–thaw cycle frequency is, of course, a highly variable element. The narrower range, at just about the freezing point is more common than the 4°F (2°C) range. The airport values in all years are higher than the urban ones, the latter showing only about two-thirds the frequency. This reflects in part the fact that mean winter temperatures in Baltimore are close to the freezing point. The seasonal variation shown in Fig. 5.26 depicts the fact that the frequency of daily freeze–thaw cycles is always higher at the airport than in the city. It also shows that the phenomenon of straddle around the freezing point in that climate starts in November and essentially ends in March.

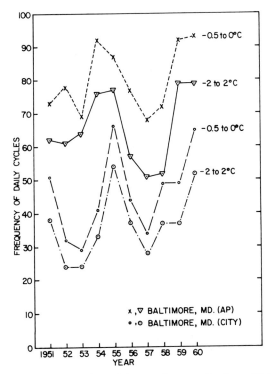

Fig. 5.26 Annual frequencies of freeze–thaw cycles for a 10-yr interval at Baltimore, Maryland, in the city and the outlying airport (from Hershfield, 1971).

References

Atwater, M. A. (1975). Thermal changes induced by urbanization and pollution. *J. Appl. Meteorol.* **14**, 1061–1071.

Böhlen, T. (1978). The influence of aerosol particles on the warming of urban and rural atmospheres. *Meteorol. Rundsch.* **31**, 87–91.

Chandler, T. J. (1965). "The Climate of London," p. 150. Hutchinson, London.

Chopra, K. P., and Pritchard, W. M. (1972). Urban shopping centers as heated islands. *Preprint, Conf. Urban Environ.; Conf. Biometeorol., 2nd.*, pp. 310–317. Am. Meteorol. Soc., Boston.

Clarke, J. F., and McElroy, J. L. (1970). Experimental studies of the nocturnal urban boundary layer. *In* "Urban Climatology," *WMO Tech. Note*, No. 108, pp. 108–112.

Clarke, J. F., and Petersen, J. T. (1972). The effect of regional climate and land use on the nocturnal heat island. *Preprint, Conf. Urban Environ.; Conf. Biometeorol., 2nd., Philadelphia,* pp. 147–152. Am. Meteorol. Soc., Boston.

Davitaya, F. F. (1958). Principles and methods of agricultural evaluation of climates. *In* "Agrometeorological Problems," pp. 62–70, Moscow.

Dettwiller, J. (1970a). Évolution séculaire du climat de Paris (Influence de l'urbanisme). *Mem. Meteorol. Natl. Paris,* No. 52, 83 pp.

Dettwiller, J. (1970b); Deep soil temperature trends and urban effects at Paris. *J. Appl. Meteorol.* **9,** 178–180.

Dettwiller, J. (1978); L'évolution séculaire de la température à Paris. *La Météorologie, VI sér.,* No. 23, 95–130.

Dronia, H. (1967); Der Städteeinfluss auf den weltweiten Temperaturtrend. *Meteorol. Abh., Berlin* **74** (4), 68 pp.

Duckworth, F. S., and Sandberg, J. S. (1954). The effect of cities upon horizontal and vertical temperature gradients. *Bull. Am. Meteorol. Soc.* **35,** 198–207.

Emonds, H. (1954). Das Bonner Stadtklima. *Arb. Rheinischen Landeskunde* No. 7, 64 pp.

Fukui, E. (1970). The recent rise of temperature in Japan. *Jpn. Progr. Climatol.,* Tokyo Univ. of Education, pp. 46–65.

Fukuoka, Y. (1980). Research in climatology by geographers in Japan. *Prof. Geographer* **32,** 224–230.

Grillo, J. N., and Spar, J. (1971). Rain–snow mesoclimatology of the New York metropolitan area. *J. Appl. Meteorol.* **10,** 56–61.

Hage, K. D. (1972). Nocturnal Temperatures in Edmonton, Alberta. *J. Appl. Meteorol.* **11,** 123–129.

Hershfield, D. M. (1971). An investigation into the frequency of freeze–thaw cycles. Ph.D. Dissertation, Dept. of Geography, Univ. of Maryland, College Park, Md., 162 pp.

Hutcheon, R. J., Johnson, R. H., Lowry, W. P., Black, C. H., and Hadley, D. (1967). Observations of the urban heat island of a small city. *Bull. Am. Meteorol. Soc.* **48,** 7–9.

Klysik, K., and Tarajowska, M. (1978). On some features of thermal structure of atmosphere ground layers above a city (authors' transl. title). *Przegl. Geofiz.* **22** (30), 43–48.

Kratzer, A. (1965). "Das Stadtklima" (2nd ed.), p. 56. Friedr. Vieweg & Sohn, Braunschweig.

Landsberg, H. E. (1950). Comfortable living depends on microclimate. *Weatherwise* **3,** 7–10.

Landsberg, H. E. (1970). Micrometeorological temperature differentiation through urbanization. *In* "Urban Climates," *WMO Tech. Note,* No. 108, pp. 129–136.

Landsberg, H. E. (1975). Atmospheric changes in a growing community. *Inst. Fluid Dynamics Appl. Math. Tech. Note,* No. BN 823, 54 pp. Univ. of Maryland, College Park, Maryland.

Landsberg, H. E., and Brush, D. A. (1980). Some observations of the Baltimore, Md., heat island. *Inst. Phys. Sci. Technol. Tech. Note.* BN-948. Univ. of Maryland, College Park, Maryland.

Lindquist, S. (1968). Studies on the local climate in Lund and its environs. *Lund Studies in Geography, Ser. A., Phys. Geograph.* No. 42, 79–93.

Linke, F. (1940). Das Klima der Groszstadt. *In* "Biologie der Groszstadt," (F. Linke and B. de Rudder, eds.), pp. 75–90. Theodor Steinkopff, Dresden.

Ludwig, F. L. (1970). Urban temperature fields. *In* "Urban Climates," *WMO Tech. Note*, No. 108, pp. 80–112.

McCormick, R. A., and Kurfis, K. R. (1966). Vertical diffusion over a city. *Q. J. Roy. Meteorol. Soc.* **92**, 392–396.

Machalek, A. (1977). Ein Beitrag zur vertikalen Temperaturverteilung über einer Groszstadt. *Ann. Meteorol.*, No. 12, 201–204.

Manley, G. (1958). On the frequency of snowfall in metropolitan England. *Q. J. Roy. Meteorol. Soc.* **84**, 70–72.

Middleton, W. E. K. (1966). "A History of The Thermometer and Its Use in Meteorology." The Johns Hopkins University Press, Baltimore, 249 pp.

Middleton, W. E. K., and Millar, F. G. (1936). Temperature profiles in Toronto. *J. Roy. Astron. Soc. Canada*, **30**, 265–272.

Mitchell, J. M. Jr. (1953). On the causes of instrumentally observed secular temperature trends. *J. Meteorol.* **10**, 244–261.

Mitchell, J. M., Jr. (1961a). The thermal climate of cities. *Symp. Air over Cities, U. S. Public Health Serv. Publ. SEC, Tech. Rept.* **A62-5**, pp. 131–143.

Mitchell, J. M., Jr. (1961b). The temperature of cities. *Weatherwise* **14**, 224–229.

Moffitt, B. J. (1972). The effects of urbanization on mean temperatures at Kew Observatory. *Weather* **27**, 121–129.

Nicholas, F. W. (1971). A synoptic climatology of Metro Washington: A mesoscale analysis of the urban heat island under selected weather conditions. M.S. Thesis, Dept. of Geog., Univ. of Maryland, 120 pp. (Unpublished.)

Nkedirim, L. C., and Truch, P. (1981). Variability of temperature fields in Calgary, Alberta. *Atmos. Environment.*

Norwine, J. R. (1972). Heat island properties of an enclosed multilevel suburban shopping center. *Preprint, Conf. Urban Environ.; Conf. Biometeorol., Philadelphia, 2nd*, pp. 139–143. Am. Meteorol. Soc., Boston. Also in: *Bull. Am. Meteorol. Soc.* **54** (1973), 637–641.

Oke, T. R. (1973). City size and the urban heat island. *Atmos. Environ.* **7**, 769–779.

Oke, T. R. (1976). The distinction between canopy and boundary layer urban heat islands. *Atmosphere* **14**, 268–277.

Oke, T. R. (1979). "Review of Urban Climatology." *WMO Tech. Note*, No. 169, 100 pp.

Oke, T. R., and East, C. (1971). The urban boundary layer in Montreal. *Boundary-Layer Meteorol.* **1**, 411–437.

Oke, T. R., and Hannell, F. G. (1970). The form of the urban heat island in Hamilton, Canada. *In* "Urban Climates," *WMO Tech. Note*, No. 108, pp. 113–126.

Reichenbächer, W. (1978). Zeitlich-räumliche Entwicklung der städtischen Wärmeinsel im Berliner Raum am 6./7. August 1974 und am 26.1 27. Februar 1975. *Beil. Berliner Wetterkarte*, 98/78, SO 23/78, Berlin, 10 pp.

Schulze, P. (1969). Die horizontale Temperaturverteilung in Groszstädten, insbesondere die West-Berlins in winterlichen Strahlungsnächten. *Meteorolog. Abh.* **91** (2), 52 pp. Dietrich Reimer, Berlin.

Summers, P. W. (1964). An urban ventilation model applied to Montreal. Ph.D. Dissertation, McGill University, Montreal. (Unpublished.)

Tyson, P. D., Dutoit, W. J. F., and Fuggle, R. F. (1972). Temperature structure above cities: Review and preliminary findings from the Johannesburg urban heat island project. *Atmos. Environ.* **6,** 533–542.

University of Alaska (1978). Research Annual Report 1977/78, (p. 8). College, Alaska 99701.

Unwin, D. J. (1980). The synoptic climatology of Birmingham's urban heat island, 1965–1974. *Weather* **35,** 43–50.

The Urban Wind Field

The changes brought about by urbanization in the local atmospheric boundary layer have a notable effect on the low-level wind. This is caused by the heat island and the change in surface roughness.

6.1 EMPIRICAL EVIDENCE FOR WIND ALTERATIONS

An early account of Kremser (1909) called attention to a decrease of wind speed in urban areas. An anemometer on top of a high school building in the suburbs of Berlin, at 32 m above the surface showed in the first decade a mean wind of 5.1 m sec^{-1}. In the next decade the formerly free terrain turned into apartment housing and the anemometer was now only 7 m above the average roof level. The mean wind speed had dropped to 3.9 m sec^{-1}, a 24 percent reduction. This is almost exactly the difference observed for the anemometers at Central Park in New York City and at La Guardia Airport. Nearly all studies show an increase in the number of calms observed in town compared with the rural areas and a notable 10–20 percent reduction in the speed of maximum winds.

In Columbia, Maryland, during the early period of urbanization, there was a gradual increase in weak winds and a decrease of strong winds. During a 6-yr period of observations 3-hourly wind-speed observations were compared with simultaneous observations at the nearby Baltimore–Washington International Airport. The Columbia wind speed was expressed as a percentage of the airport wind speed. These observations were grouped in three classes:

(1) Columbia wind speed was less than 70 percent of that observed at the airport;

(2) 70–99 percent of airport speed; and

(3) the speed was as high or higher in Columbia than at the airport.

The results are shown in Fig. 6.1. In 1969 wind speeds in Columbia were higher than at the airport in 25 percent of the cases, but by 1974 this class had dropped to 14 percent of the observations. In the same interval the wind speeds, which were less than 70 percent in the newly urbanized area of those recorded at the airport, had increased from 43 to 65 percent.

A very interesting case of a long-term trend in wind speed caused by urbanization has been published by Rubinshtein (1979). It showed for the growing town of Gantsevitchi in White Russia a monotonous drop from mean annual velocities of 3.9 m sec^{-1} in 1945 to 2.5 m sec^{-1} in 1971, as shown in Fig. 6.2. This is a reduction of about 36 percent.

Another excellent example of the urban effect on wind speed has been reported by Zanella (1976) for Parma located in the Po plain of northern Italy. With a progressively expanding urban area the annual number of calms is 55 percent of the observations, but only 48 percent at the airport. In the winter season these percentages go up to 82 percent in the city versus 64 percent at the airport. In that season and in spring, wind speeds are at all hours of the day lower in the city than at the airport. In summer and autumn the difference vanishes only in the early evening hours. Most revealing is the change from decade to decade shown in Table 6.1. The total speed

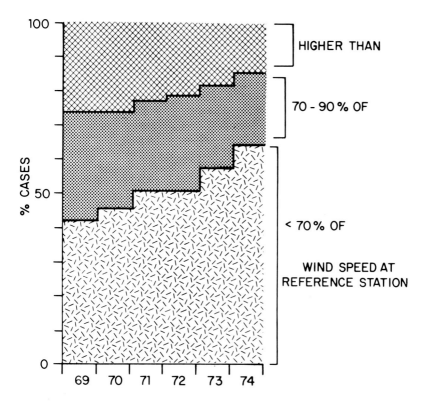

Fig. 6.1 Change in wind speed classes during rapid growth era of Columbia, Maryland.

reduction in a quarter century has been 39 percent. All directions appeared to be equally reduced.

The low winter wind speeds in Parma are attributable to the regional climate. In other regions the urban wind speeds are materially affected by the vegetation. Dirmhirn and Sauberer (1959) indicated that in Vienna when the deciduous trees were in leaf, urban wind speeds dropped 20–30 percent. The same effect was observed by Frederick (1961) in Nashville, Tennessee. There 24-hr wind movement was measured by totaling anemometers on 32 utility posts, about 10-m above the surface. The ratio of the wind movement at these urban stations was compared to that at the airport. This analysis showed that during summer in an area of numerous de-

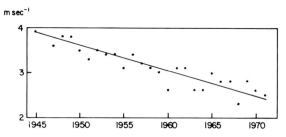

Fig. 6.2 Time series of mean annual wind speed in the growing city of Gansevitchi, U. S. S. R. (based on data by Rubinshtein, 1979).

ciduous trees, spread throughout the city, wind speeds were reduced 20–30 percent compared to sites not affected by trees. After defoliation wind speeds at the sites with trees increased by 25–40 percent. On the whole, reductions of 30–60 percent in wind speed at the various urban sites were noted.

A comparative set of observations above roof level was made in Vienna (Steinhauser *et al.*, 1959). One anemometer was installed on the roof of the Technical University in the densely built-up center of the city, the other at the edge of the city, in a parklike setting, on an observatory. Ratios of wind speeds at the two sites for various wind directions and speed classes were compiled. The results are shown in Table 6.2. The table shows a number of effects. For both listed wind directions the greater speeds are less affected by the urban area than the weaker ones. In summer the suburban site shows for west winds a higher ratio than in winter. The cause is the already discussed leaf effect, with many deciduous trees near the observatory. For the SE winds the ratios for the stronger winter winds and all speed classes in summer are reversed: the city wind speeds are

TABLE 6.1

Changes in Mean Wind Speeds in Parma, Italy, in Three Consecutive Decades[a]

Interval	January	April	July	October	Year
1938–1949	0.5	1.8	1.8	1.0	1.3
1950–1961	0.5	1.4	1.4	0.7	1.0
1962–1973	0.3	1.0	1.3	0.6	0.8

[a] Wind speeds (m sec^{-1}). Adapted from Zanella (1976).

TABLE 6.2

Wind Speed Ratios, Center–Suburb, in Vienna, Austria[a]

Wind direction	Season	Wind speed (km hr^{-1})			
		5 (1.4)[b]	15 (4.2)	25 (6.9)	35 (9.7)
W	Winter	0.5	0.6	0.7	0.77
	Summer	0.75	0.82	0.9	0.85
SE	Winter	0.5	0.8	1.05	
	Summer	1.0	1.25	1.25	

[a] After Steinhauser *et al.* (1959).
[b] Values in parentheses in m sec^{-1}.

higher than suburban values. This is produced by higher roughness in the wind-fetch area on the approach to the observatory.

An early observation has been low-level convergence of air flow into urban areas. This is a logical consequence of the heat island, which creates an unstable vertical lapse rate of temperature and thus induces a rising air current. More details on that will be presented in Section 6.2. Observations of inflow of air to the urban center are easiest at night and an example of such flow is shown in Fig. 6.3 (Stummer, 1939). Usually, even with a well-developed nocturnal heat island, concentric convergence is a rare case. The irregularity of urban structure and surrounding terrain dissimilarities prevent that. There are a few well-documented cases of centripetal urban air flow in the literature. One deserves to be singled out because of the ingenious method of observation. This has been reported by Okita (1960) from the city of Asahikawa (Hokkaido, Japan). This place of about 190,000 inhabitants has in winter frequent radiation fog at temperatures much below freezing. On some of these occasions heavy rime is formed on the windward side of trees by impinging of supercooled droplets. From these Okita was able to determine the wind direction at many localities around the city. On days with a minimum temperature difference (urban–rural) of about 4°C, implying a maximum heat-island difference during the night of 8°C, the convergence was almost perfect.

The nocturnal convergence into the urban areas is the reason for the observation that wind speeds at night in cities are not as much weakened as in daytime and occasionally can be stronger than in the

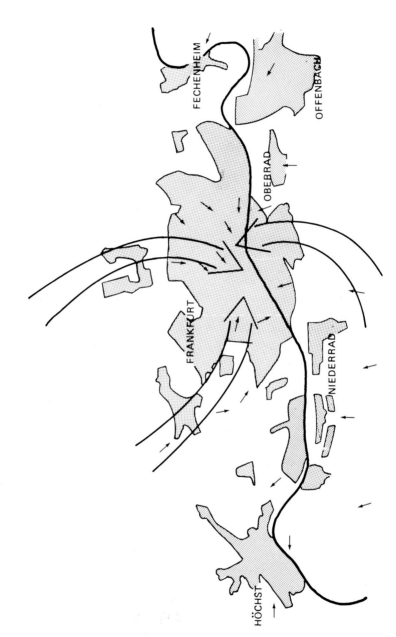

Fig. 6.3 Nocturnal wind convergence on calm nights into Frankfurt am Main, Germany (adapted from Stummer, 1939).

country. That happens in particular when under nearly calm conditions a strong rural inversion builds up while the lower layer in the city remains unstably stratified. Chandler (1965) shows this well for London in a comparison of winds at 0100 hr GMT and 1300 hr GMT in the city center and the airport, as given in Table 6.3.

Lee (1979) clearly showed for London that the small nocturnal wind speed difference in and outside the city is a direct function of the frequency and intensity of the rural inversions. And although in daytime wind speeds are reduced on an average by 30 percent in the city, the average nighttime reduction is only 20 percent. In smaller towns the day–night differential is not nearly as pronounced.

It is also essential to understand that the wind flow at night into the urban area is not a steady one. In large urban areas the isotherms of the heat island, if it has indeed a single core, are not equidistantly spaced. They are usually crowded near the edge of the built-up area. Meteorologically this is analogous to a cold front where temperatures change abruptly over short distances. This characteristic of the heat island can lead to sharp pulses of cooler country air invading the city at night, often with a notable wind gust. As early as 1925 Schmauss presented a case of that type from Munich and called attention to these miniature cold fronts. Figure 6.4 shows two such invasions in one night. In daytime the wind flow into the urban area is not as readily discernible and in many cases can only be noted as a deformation of the general wind field. Occasionally on hot summer

TABLE 6.3

Average Seasonal Wind Speeds[a] at the Airport and the City of London (1961–1962) at Different Hours

Season	0100 hours		1300 hours	
	Airport	City difference (higher)	Airport	City difference (lower)
Winter	2.5	0.4	3.1	−0.4
Spring	2.2	0.1	3.1	−1.2
Summer	2.0	0.6	2.7	−0.7
Autumn	2.1	0.2	2.6	−0.6
Year	2.2	0.3	2.9	−0.7

[a] Values in m sec^{-1}.
[b] After Chandler (1965).

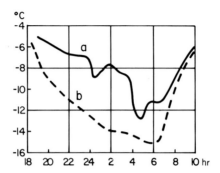

Fig. 6.4 Nocturnal march of temperature in Munich, Germany (solid line, a) and at a suburban station (dashed line, b) during winter night, showing in "a" two frontlike invasions into the city (after Schmauss, (1925).

afternoons with weak gradient winds the convergence can be directly observed (Fig. 6.5).

Berg (1947) made the attempt to use the elementary relations between temperature, pressure, and wind to estimate the speed of the country breeze.[1] He figured that there is a 5°C ΔT_{u-r}, decreasing to 1°C at 500 m and vanishing at 1000 m. Assuming a 10-km distance for this temperature difference leads to a pressure difference of 0.07 mbar (7 Pa). Such a small pressure difference is not measurable with ordinary equipment and has indeed never been observed in urban climate studies, but the calculated resulting surface wind is about 3 m sec^{-1}, a value entirely compatible with the observations.

One additional fact about the surface near winds in urban areas is readily noted in anemograms. The wind turbulence shows a distinct increase in the gusts with higher frequencies compared to open areas. A study of the power spectra of wind speeds at a large apartment complex and a nearby airport (Badger, 1973) showed that at the airport, where the wind had free and rather unobstructed fetch, most of the power is in the low frequencies (>1 min) and the spectrum is very close to a red-noise spectrum. This means that the turbulent eddies are large. In the apartment complex there is a notable shift in the power spectrum to higher frequencies. At that end of the spectrum there are notable departures from red noise and small eddies clearly dominate in that rough urban environment.

---

[1] We use the term "country breeze" in the same sense as those of other secondary circulations, such as mountain breeze or sea breeze, each indicating, as in general meteorology, the direction the wind is coming from.

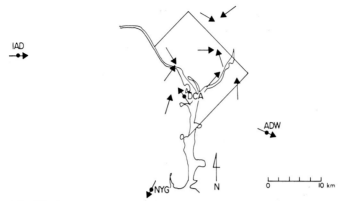

Fig. 6.5 Wind convergence on a hot summer afternoon in Washington D.C. IAD, Dulles Intl. Airport; DCA, Washington Natl. Airport; ADW, Andrews Airbase; NYG, Quantico Marine Base.

6.2 VERTICAL WIND STRUCTURE IN URBAN AREAS

Several techniques have been complementarily used to gauge wind conditions over cities. The simplest are instrumented towers, which have increased in number in cities for broadcast and telecommunication purposes. But many years of observations are available from Paris where Alexandre Gustave Eiffel started wind observations on the 300-m-high tower designed by him in 1890. Television towers are usually less than 200 m in height but if equipped with anemometers at various levels they can contribute useful information. Interpretation of observations thus obtained may require corrections if the wind flows through the tower structure before striking the anemometer (Izumi and Barad, 1970).

Two procedures using balloons have been employed. The vertical profile is obtained by ordinary pilot balloon observations. If simultaneous values are obtained at various points around a city, convergence and divergence can be determined. Use of constant-volume balloons, so-called tetroons, has been made to obtain trajectories of flow over urban areas. These tetroons fly at a level of constant density. They carry a transponder and are tracked by radar. From their track and flight level the vertical component of air motion can be determined. In addition, observations from low-flying aircraft can contribute to the wind information over urban areas. Yet, all these airborne observations have the drawback of being essentially sporadic. But they can reach into higher air layers than the towers, which furnish continuous data.

The vertical profiles in low levels show one phenomenon clearly. The wind, which is notably retarded below the level of the buildings, shows usually a minor speed maximum just above the mean roof level, that can be designated as a roof-top jetlet. The increase is only about 1 m sec^{-1} on an average. However, in very strong winds, for example in typhoons, gust speeds may be much higher than surface winds and can dominate the total low-level wind profile. Observations on a 333-m-high tower in Tokyo by Arakawa and Tsutsumi (1967) showed for winds ≥ 30 m sec^{-1} that a level above the roof level of 30-m, gust speeds were commonly 10 m sec^{-1} higher than just below the roof niveau.

Tower observations have long been used to characterize air flow in the planetary boundary layer. Obviously the massive obstacles introduce major perturbations, but Munn (1970) has cogently pointed out how important it is to know this urban flow for engineering applications and for pollutant-dispersal estimates. Over smooth surfaces meteorologists have adopted the theory of flow in fluids, first developed by Ekman (1905) in a classical treatise. By analogy this was transferred to the atmosphere. The theory states that under ideal conditions, the surface influence causes the vectors of motion in successive layers, up to the layer where friction ceases, to have an effect on the wind: to form an equiangular (logarithmic) spiral. The wind, unimpeded by friction is called the geostrophic wind, represented by

$$V_G = -\frac{1}{\rho f}\frac{\partial p}{\partial x} \qquad (6.1)$$

where

V_G　geostrophic wind
ρ　atmospheric density
f　Coriolis parameter ($2\Omega \sin \phi$; Ω, angular velocity; ϕ, latitude)
$\partial p/\partial x$　horizontal pressure gradient from higher to lower pressure

The geostrophic wind blows parallel to the isobars at that level. The nongeostrophic wind components below, in the ideal case, make an angle of 45° with the tangents to the spiral. One can express for averaged wind vectors the relation between the geostrophic wind and the wind in the layers below as

$$\bar{v}_z - \bar{V}_G = \frac{1}{f} \frac{d}{dz}\left(\frac{\tau_z}{\rho}\right) \tag{6.2}$$

where

$\bar{v}_z$ average wind vector at any level z below the geostrophic wind level

τ_z shear stress

ρ atmospheric density

The shear stress, also termed Reynolds stress, can be determined from the turbulent components of wind speed, i.e., the departures from the average in the three components u', v', w'. For convenience a parameter called friction velocity u_* has been introduced. It equals $(\tau/\rho)^{1/2}$ and is a function of the mean horizontal and vertical wind fluctuations

$$\frac{\tau}{\rho} = u_*^2 = \langle -u'w' \rangle \tag{6.3}$$

These wind fluctuations can be measured on towers.

In a neutrally stratified atmosphere, i.e., with an adiabatic vertical temperature lapse rate of 1°C/100 m, aerodynamic theory then permits us to derive a vertical wind profile (Taylor, 1952)

$$\bar{u}_z = \frac{u_*}{k} \ln\left(\frac{z}{z_0}\right) \tag{6.4}$$

where

k von Kármán's constant (~ 0.4)

z height

z_0 friction height or roughness parameter

This is usually referred to as the logarithmic wind profile. In any environment the value of z_0 can be determined by wind observations at two heights:

$$\ln z_0 = \frac{\bar{u}_2 \ln z_1 - \bar{u}_1 \ln z_2}{\bar{u}_2 - \bar{u}_1} \tag{6.5}$$

where the subscripts refer to the values at the two heights.

Table 6.4 shows a number of values reported in the literature for urban areas. As is readily seen, the values scatter widely, by almost an order of magnitude. This is the result of varying roughness upwind of the measuring tower, the length of fetch, and the degree to which the neutral stability condition was fulfilled. On the whole, however, there is no doubt that the urban roughness parameter is generally an order of magnitude larger than over agriculturally used land. From measurements made in the Paris, France area, including the values from the Eiffel tower (Dettwiller, 1969), one can derive values ranging from 2 to 5 m.

A theoretical approach for estimating the roughness parameter has been suggested by Lettau (1969, 1970) using the similarity principle. According to this scheme, z_0 can be predicted if the effective heights of obstacles are known, the subtended area encountered by the wind, and the area covered can be ascertained. In that case

$$z_0 = \frac{1}{2} \frac{Ha}{A} \qquad (6.6)$$

where

H height of obstacle (m)
a silhouette area encountered (m²)
A area covered by obstacles (m²)

and the factor $\frac{1}{2}$ approximates the average drag coefficient of roughness elements. Estimates of roughness parameters using this

TABLE 6.4

Measured Roughness Parameters z_0 in Urban Areas

Locality	z_0 (m)	Source
Kiev, U.S.S.R.	4.5	Ariel and Kliuchnikova (1960)
Fort Wayne, Indiana	3.0	Csanady et al. (1968)
Minneapolis, Minnesota	2.0	Deland and Binkowski (1966)
Tokyo, Japan	1.7	Yamamoto and Shimanuti (1964)
Liverpool, England	1.2	Jones et al. (1971)
Austin, Texas	0.4–2.4	Peschier (1973)
Reading, England	0.7	Marsh (1969)
Cambridge, Massachusetts	0.5–2	Dobbins (1977)
Columbia, Maryland	0.7	Landsberg (1979)

TABLE 6.5

**Aerodynamic Roughness Based on Building
Characteristics in Urban Areas**[a]

	Building type		
Parameter	Low	Medium	High
Height (m)	4	20	100
Silhouette (m²)	50	560	4000
Built-up area (m²)	2000	8000	20,000
Calculated z_0 (m)	0.5	0.7	10

[a] According to Lettau (1970).

scheme are given in Table 6.5. The values obtained this way seem to yield realistic numbers that are quite comparable to the measured figures. Lettau uses the logarithmic wind profile in the following form for the Ekman boundary layer, using logarithms to base 10:

$$\bar{u}_z = 5.5 c V_G \log_{10}(1 + z/z_0) \tag{6.7}$$

where

$\bar{u}_z$ mean wind at height z
c the drag coefficient[2]

Lettau and others have observed and calculated Ekman spirals for various drag conditions. The observed values only approximate the theoretical model, which has been used to simulate urban conditions for wind tunnel experiments. Dobbins (1977) offers measured values from observations made in Cambridge, Massachusetts. He estimated from his data that the atmospheric boundary layer ranges from 10 to 40 times the roughness height z_0. A typical hodograph from his work is shown in Fig. 6.6. The hodograph is the envelope of the wind vectors at various heights above the surface.

In other representations of the vertical wind profile, a simple power law was used; this is especially notable in engineering litera-

[2] This is related to the nondimensional Rossby number Ro $= V_G/z_0 f$, where f is the Coriolis parameter. Csanady (1967) gave this relation: Ro $= V_G/z_0 = 0.4((1/c) - 115)^{1/2} + 1.15 \ln(1/c) - 1.52$.

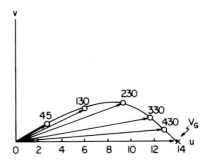

Fig. 6.6 Hodograph of wind field over urban area of Cambridge, Massachusetts. Numbers on wind-vector envelope are heights in meters; abscissa shows wind speed in meters per second (adapted from Dobbins, 1977).

ture. Two forms are common:

$$\bar{u}_z = u_1\left(\frac{z}{z_1}\right)^{\alpha} \tag{6.8}$$

and

$$\bar{u}_z = V_G\left(\frac{z}{z_G}\right)^{\alpha'} \tag{6.9}$$

where α is an exponent representing all the friction factors. The version using the geostrophic wind was introduced by Davenport (1965). This author gives the value of the exponent α' and the average level of the gradient wind as shown in Table 6.6.

The average vertical wind distribution in percent of the gradient wind is depicted in Fig. 6.7. Obviously such an idealized picture is only valid for a neutral stratification, and wide deviations, depending on the synoptic situation and time of day, are observed in practice.

TABLE 6.6

Measurements of the Vertical Wind Profile for Different Surface Roughnesses[a]

Terrain characteristic	Exponent of power law	Gradient wind level (m)
Open country, flat	0.16	270
Suburban settlement	0.28	390
Inner cities	0.40	420

[a] Davenport (1965).

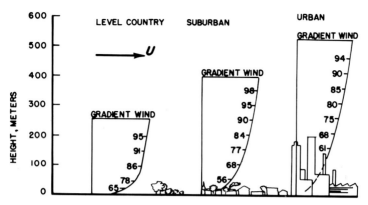

Fig. 6.7 Vertical wind speeds, in percent of the gradient wind at various heights over terrain of different roughness (after Davenport, 1965).

The exponent in Eq. (6.8) has been determined empirically to vary in open country between 0.07 and 0.25. A commonly used value is 0.14. Borisenko and Zavarina (1968) have determined the dependence of the value of α in the lowest 100 m of the atmosphere on the roughness parameter z_0. An excerpt from their table, applicable to suburban and urban environments is shown in Table 6.7. The relation between these two variables is fairly linear in this range and can be approximated by $\alpha = 0.12z_0 + 0.18$. The value for the 2-m

TABLE 6.7

Relation between Roughness Parameter z_0 and Exponent α in Power Law of Vertical Wind Profile in the Lowest 100 Meters[a]

z_0 (m)	α
0.1	0.18
0.2	0.21
0.5	0.25
1.0	0.31
2.0	0.42

[a] Adapted from Borisenko and Zavarina (1967).

roughness parameter compares well with an α value of 0.44 measured over the most densely built-up area of Columbia, Maryland. Measurements made by Jackson (1978) in the windy city of Wellington, New Zealand, between 10 and 70 m above ground, yield an average α of 0.5. For northerly winds z_0 was determined at 3.8 $\pm$ 0.6 m. The greatest turbulence was below the average roof height at the 10-m level. It rapidly decreased to the 70-m height.

A great deal of information not only on wind flow over urban areas but on the vertical component of wind motion has been obtained from tetroon flights. Much of this information has been obtained over New York City (Hass *et al.*, 1967; Angell *et al.*, 1968; Druyan, 1968). These flights were made at about 300 m and their main objective was the collection of facts about air trajectories for travel of air pollutants. Obviously, because of the coastal location one has to be careful not to generalize the results too much as characteristic for urban climates. Yet a few points stand out. Over the center of the urban area in Manhattan the tetroon drift was only about $\frac{2}{3}$ of the geostrophic wind. The vertical motion obviously was greatly influenced by the atmospheric stability. Early morning flights were far more level than midday flights when temperature lapse rates were large. In midday the heat-island effect caused notable vertical lift in the balloon flights over the city, but when crossing the Hudson River, where cool water stabilized the air, the balloons returned to lower levels. This is very well demonstrated in Fig. 6.8, which shows the vertical track of a flight near noon across Manhattan Island. The flight levels over water surfaces are low, but over the built-up areas they float several hundred meters higher. The jump

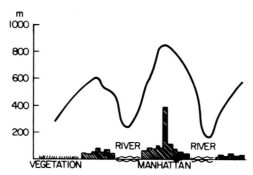

Fig. 6.8 Vertical track of a tetroon on a midday flight across Manhattan Island, New York (after Hass *et al.*, 1967).

over the Empire State Building (381 m) is particularly spectacular, where the balloon climbed above 800 m.

Vertical velocities determined from the tetroon flights over New York City were twice as large in midday than either in the early morning or evening. Between 0900 and 1500 hr local time, at a height of 300–500 m, it exceeded 1 m sec^{-1} in 15 percent of the observations and 0.4 m sec^{-1} for 50 percent of the observations. The convection currents over cities, resulting in "bumpiness," have been well-known since the early days of manned flight.

Another fact noted is that the tetroons have a tendency to move toward lower pressure on the days when the sea breeze did not interfere with the flights. The tetroon observations were confirmed and extended by the work of Ackerman (1972, 1977) and Auer (1975) over St. Louis in the METROMEX study. The information gathered there is likely to be more representative than the New York data because of the absence of topographic disturbances. The analysis by Ackerman was based on pilot balloon observations to which Auer added wind information obtained by aircraft tracked by Doppler radar.

These studies clearly showed convergence over the urban area and slightly cyclonically curved trajectories. Obviously, the wind field distortion was most notable when regional winds were weak but still observable even in strong winds. The divergence is calculated from the wind field by measuring the increase in area of the horizontal winds subtended in their passage over the city during a given time interval

$$D = \frac{1}{A_0} \frac{\Delta A}{\Delta t} \tag{6.10}$$

where

D divergence (if negative: convergence)
A_0 initial arc
ΔA difference between A and distortion of winds in time lapse Δt

Of course, the procedures permit only limited samples, but in her work Ackerman was able to use data for 21 undisturbed summer afternoons in the lowest 1500 m above St. Louis. These data should reflect the urban contribution to the wind field quite plausibly. In the lowest layers there is mass convergence into the city with D values of -1×10^{-4} sec^{-1}, reaching values up to $\sim 1.7 \times 10^{-4}$ sec^{-1}. At

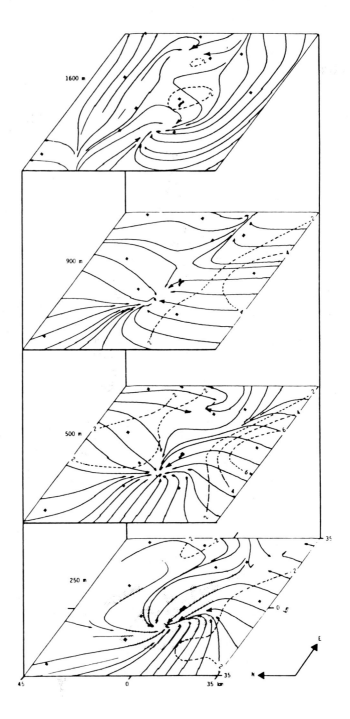

about 300 m the convergence changes to divergence. Calculations of vertical motion yielded mean values of $\bar{w} = 0.07$ m sec^{-1} to maximally 0.3 m sec^{-1}. The afternoon values showed convergence into the city in about 60 percent of all observations, with the greatest disturbance shown in the early afternoon at the time of the maximum temperature. Figure 6.9 shows the urban wind perturbation over St. Louis for a typical case at midday.

6.3 URBAN WIND FIELD AND OTHER MESOSCALE CIRCULATIONS

The urban wind field is rarely simple. Even when the synoptic situation is least complicated, with clear skies and weak winds in the center of an extended high-pressure area, small differences in local topography will cause irregular air flows. Thus the broad generalizations that have been introduced into the literature are at best guideposts to the phenomena. Even in the case of St. Louis, where the terrain has relatively less influence than in case of other cities, one can only take the data presented by Dannevik, *et al.* (1974) as order-of-magnitude estimates, shown in Table 6.8.

Similarly, the values obtained for the equally simple terrain of Oklahoma City by Angell *et al.* (1973) in strong winds (≥ 13 m sec^{-1}) still raise notable questions. The daytime values when the heat-island effect is small, measured by tetroon and on a 460-m-tall television tower, suggest that the city acts simply as an obstacle to the air flow. Thus there is a tendency for streaming around the city and a vertical component introduced by the barrier effect. The rising current in daytime is about $0.3 - 0.4$ m sec^{-1} close to values obtained in other experiments. A 5° turning toward the low-pressure side was noted and the surface stress estimated at 2 dynes cm^{-2}.

But when we turn from these very simple settings, many complexities arise. It should be emphasized here that cities usually have

---

Fig. 6.9 Wind perturbation field at noon on a summer day over St. Louis, showing the convergence field by streamlines (solid curves), perturbation isotachs in meters per second (dashed curves) at various levels above the city (heights are given in meters above mean sea level) (from Ackerman, 1972).

TABLE 6.8

Estimates of the St. Louis, Missouri, Heat-Island Circulation Elements[a]

Element	General magnitude
ΔT_{u-r}	$\geq 2°C$
V_G (900 cPa)	≥ 5 m sec^{-1}
$\bar{u}$	2 m sec^{-1}
$\overline{w}$	0.3 m sec^{-1}
Diameter of:	
surface inflow	30 km
updraft	7 km
Depth of circulation	1 km

[a] From Dannevik et al. (1974).

arisen because of some terrain features favorable to one or more human activities. Most common are locations favoring various forms of traffic. Thus, river valleys, bays, and shore locations on lakes and on the sea are preferred spots for urban areas. These landscapes develop invariably their own secondary circulations, such as land and sea breezes, and mountain and valley breezes. In some instances both of these circulations can be present. They interact with the urban-induced circulation and may, at times, dominate it.

A prize example of the latter variety is the Los Angeles basin, where land and sea breezes reinforce mountain and valley breezes. They cause inland flow in daytime and seaward flow at night. Angell et al. (1966) were the first to apply their tetroon techniques to the air flow in that area. Flights at 300 m clearly showed the reversal of flow from land to sea and sea to land as a function of time of day. Vertical velocities were again of the order of magnitude of 0.5 m sec^{-1} over land and, in the stable air over water, about 0.2 m sec^{-1} at sea. The land and sea breeze seemed not to exceed 2 m sec^{-1}. Although this study did not specifically refer to any mountain-induced circulation, its existence in that area is well known. As far as one can tell, the urban influence is completely masked. It is regrettable that, at least in the air pollution literature, the completely uncharacteristic patterns of the Los Angeles basin dominated urban meteorology.

As work on other metropolitan areas progressed, better information on urban effects on the secondary circulations became available. It is not too surprising that one finds that each city has its own patterns. The magnitude of urban influence can be minor or major. In some cases notable microclimatic differences due to cold-air drainage are clearly marked. Patterson and Hage (1979) recorded an October observation in Edmonton (Alberta) in the North Saskatchewan River valley. A steep bank, with a 50-m elevation difference caused a slope flow of 0.4 m sec^{-1} on a clear night to develop an intense inversion, with a vertical gradient of $\sim 12°C/100$ m. Less than a kilometer away in the city center the temperature inversion had only a gradient of $\sim 2°C/100$ m.

In the same area, Calgary (Alberta) on the flood plains of the Bow and Elbow rivers, 110 km east of the Rocky Mountains, developed similar conditions on a January night, as reported by Nkedirim (1980). A 200-m elevation difference in the town site led to cold-air drainage and to a reduction of the rural–urban temperature contrast. Although low temperatures of between $- 8$ and $- 18°C$ were noted in the area, the difference from city center to city edge was only 4°C. Nkedirim commented that Summer's model (see p. 95) would predict a difference ΔT_{u-r} of about 6.6°C and he concludes that the cold-air drainage strengthens stability and reduces the effect of anthropogenic heat production on the urban heat island.

A rather opposite effect has been observed by Goldreich (1979a,b) in the Bezuidenhout Valley of South Africa. There nocturnal katabatic winds develop on clear winter nights, with site elevation differences of about 200 m. The slight breezes are generally below 1 m sec^{-1}. The city of Johannesburg apparently warms these up, and in the Bezuidenhout Valley temperatures are nocturnally 0.5–1°C higher than in air flowing down the tributary valleys. Unfortunately, no sustained observations of the effects of an urban area on a well-developed mountain and valley breeze system, occurring with high frequency is as yet available.

There is far more systematic information on the effect of urban areas on land and sea breezes, much of it due to Bornstein (1975) and collaborators for New York City, which shall be discussed later in greater detail. Auliciems (1979) has reported on some conditions for Brisbane (Queensland). There a weak land breeze from the southwest, of 2–3 m sec^{-1} develops in winter, probably weakened by a 5°C heat island. In summer a more vigorous sea breeze of 5–6 m sec^{-1} develops and is assumed, in conjunction with a weak nocturnal land breeze, to contribute to the recirculation of pollutants produced by the urban area.

Colacino and Dell'Osso (1978), made some studies of the development of the sea-breeze circulation in the region of Rome, Italy. Their observations include measurements of infrared surface temperatures on a day near solstice. At that time the sea-surface temperature was between 19 and 21°C. At 0600 hr, surface temperatures on land were 12°C in the country and 18°C in Rome. By 1330 hr the country surface temperatures had risen to 36°C and to 50°C in Rome. The maximum air temperatures were 21.2°C at sea, 26°C at a rural station, and 28°C in Rome. At night, from 2100 to 0600 hr, there was hardly any air motion. No well-developed land breeze existed, but at 0800 the sea breeze started to cross the coast at 4 m sec^{-1}. By 1500 hr it reached a peak velocity of 8 m sec^{-1}, a vigorous sea-breeze response to the 30° infrared temperature contrast between the sea and the city.

Bornstein and collaborators treat this penetration of the sea breeze in the New York area, by case studies, as a frontal phenomenon. Credit is properly given to Koschmieder (1935), who first recognized that the advance of the sea breeze resembles a miniature front. In some respect the work of Bornstein *et al.* (1978, 1979) yields results on the city influence that correspond to the obstacle effect observed by Angell *et al.* (1973) for Oklahoma City. Isochrone charts of the movement of the sea-breeze front reveal a frictional retardation by the built-up areas of New York City. Figure 6.10 shows very clearly the crowding of the isochrones over the built-up area (Fontana and Bornstein, 1979). From tetroon flights Anderson and Bornstein (1979) calculated the steepening of the sea-breeze front. They state: "Results show a dramatic steepening of frontal slope over the central urban area." In rural areas, the frontal slope $\Delta z/\Delta x$ was usually in the neighborhood of 1:100, a value not much different from that observed for many cold fronts. But urban values as low as 1:17 were noted in the city area. Six values in Anderson and Bornstein's data set were ≤1:71, which is ascribed to retardation at the surface and acceleration aloft. Vertical velocities of 0.65 m sec^{-1} were noted from these tetroon flights. The frictional retardation was also noted for the advance of the lake-breeze front in Chicago (Landsberg, 1958), where on many occasions this front showed a typical "nose", i.e., it advanced at 50 m, just above roof height, earlier over the built-up area than at the surface.

The effect of urban friction on fronts was first discussed by Belger (1940). For a small sample of four cases of migrating rain fronts, he

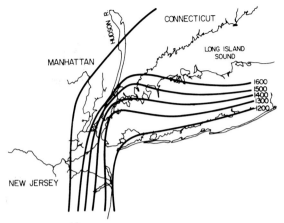

Fig. 6.10 Isochrones of a sea-breeze front in the New York City area, showing the retardation in the built-up area (after Fontana and Bornstein, 1979).

showed a retardation of 25 percent in speed over Berlin, Germany. This metropolis deformed the fronts and Belger speculated that this caused a local increase in rainfall by the lingering of the rain-producing discontinuity. We will pick up this theme again in Chapter 8. Loose and Bornstein (1977) also address the problem of frontal slowdown over New York City. From a number of cases with fronts moving from various directions they find an even more pronounced retardation of frontal movement in the urban area than Belger. Their value is about 50 percent, almost uniformly for fronts whose movement over rural areas ranged from 4 to 14 m sec^{-1}. They note that these observations agree qualitatively but not quantitatively with Bornstein's 1975 URBMET model which will be discussed in Chapter 7.

References

Ackerman, B. (1972). Winds in the Ekman layer over St. Louis. *Conf. Urban Environ.; Conf. Biometeorol., Philadelphia, 2nd,* pp. 22–27. Preprints, Am. Meteorol. Soc., Boston.

Ackerman, B. (1977). Mesoscale wind fields over St. Louis. *Conf. Inadvertent Planned Weather Modification, Champaign-Urbana, 6th,* pp. 5–8. Preprints, Am. Meteorol. Soc., Boston.

Anderson, S. F., and Bornstein, R. D. (1979). Effect of New York City on the horizontal and vertical structure of sea breeze fronts (Final Report), Vol. II, 49 pp. Dept. of Meteorology, San Jose State Univ., San Jose, California.

Angell, J. K., Pack, D. H., Holzworth, G. C., and Dickson, C. R. (1966). Tetroon trajectories in an urban atmosphere. *J. Appl. Meteorol.* **5,** 565–572.

Angell, J. K. Pack, D. H., Hass, W. A., and Hoecker, W. H. (1968). Tetroon flights over New York City. *Weather* **23,** 184–191.

Angell, J. K., Hoecker, W. H., Dickson, C. R., and Pack, D. H. (1973). Urban influence on a strong daytime air flow as determined from tetroon flights. *J. Appl. Meteorol.* **12,** 924–936.

Arakawa, H., and Tsutsumi, K. (1967). Strong gusts in the lowest 250-m layer over the city of Tokyo. *J. Appl. Meteorol.* **6,** 848–851.

Ariel, M. Z., and Kliuchnikova, L. A. (1960). Wind over a city. *Tr. Glav. Geofiz. Obs.* **94,** 29–32 (Eng. Transl. by I. A. Donehoo, U. S. Weather Bur., Washington, D. C., 1961, 6 pp.).

Auer, Jr., A. H. (1975). Calculations of divergence from the wind field over an urban area. *U. S. Natl. Conf. Wind Eng. Res., Ft. Collins, Colorado, 2nd,* I-20-1 to I-20-3 (J. E. Cermak, ed.).

Auliciems, A. (1979). Spatial, temporal and human dimensions of air pollution in Brisbane, 91 pp. Publications Dept. of Geography, Univ. of Queensland, St. Lucia.

Badger, N. K. (1973). Spectral characteristics of micropulsations of pressure in the boundary layer, 51 pp. M. S. Thesis, Univ. of Maryland, College Park.

Belger, W. (1940). Der Groszstadteinfluss auf nichtstationäre Regenfronten und ein Beitrag zur Bildung lokaler Wärmegewitter, 69 pp. Dissertation, Univ. Berlin, Orthen, Köln.

Berg, H. (1947). "Einfübrung in die Bioklimatologie," p. 69. H. Bouvier & Co., Bonn.

Borisenko, M. M., and Zavarina, M. V. (1961). Vertical profiles of wind velocities according to measurements on high meteorological towers (translated title). *Tr. Gl. Geofiz. Obs.,* No. 210.

Bornstein, R. D. (1975). The two dimensional URBMET urban boundary layer model. *J. Appl. Meteorol.* **14,** 1459–1477.

Bornstein, R. D., Fontana, P. H., and Thompson, W. T. (1978). Frictional retardation of sea breeze front penetration in New York City and its implication for pollutant distribution. *WMO Symp. Boundary Layer Phys. Applied to Specific Problems in Air Pollution, Norrköping, Sweden,* 8 pp.

Bornstein, R. D., Fontana, P. H., and Thompson, W. T. (1979). Effect of sea breeze and synoptic front passages on sulfur dioxide concentrations in New York City. *Symp. Turbulence, Diffusion and Air Pollution, Reno, 4th,* pp. 429–434. Am. Meteorol. Ser., Boston.

Chandler, T. J. (1965). "The Climate of London," p. 71. Hutchison, London.

Colacino, M., and Dell'Osso, L. (1978). The local atmospheric circulation in the Rome area: Surface observations. *Boundary Layer Meteorol.* **14,** 133–151.

Csanady, G. T. (1967). On the "resistance law" of a turbulent Ekman layer. *J. Atmos. Sci.* **24,** 467–471.

Csanady, G. T., Hilst, G. R., and Bowne, N. E. (1968). Turbulent diffusion from a cross-line source in shear flow at Fort Wayne, Indiana. *Atmos. Environ.* **2,** 273–292.

Dannevik, W. P., Frisella, S., Hwang, H. J., and Husar, R. B. (1974). Analysis and characterization of heat island-induced circulation during the 1973 EPA characterization study, St. Louis, Missouri. *Conf. Weather Forecasting Anal., 5th.* Preprints, Am. Meteorol. Soc., Boston.

Davenport, A. G. (1965). The relationship of wind structure to wind loading. *Proc. Conf. Wind Effects on Structures,* Vol. I, pp. 53–102. Natl. Phys. Lab., H. M. Stationery Office, London.

Deland, R. J., and Binkowski, F. S. (1966). Comparison of wind at 500 feet over Minneapolis and Louisville with geostrophic wind. *J. Air Pollut. Control Assoc.* **16,** 407–411.

Dettwiller, J. (1969). Le vent au sommet de la tour Eiffel. *Monogr. Meteorol. Natl.,* No. 64, 28 pp.

Dirmhirn, I., and Sauberer, F. (1959). Das Strassenklima von Wien. *In* "Klima und Bioklima von Wien," (Steinhauser, F., Eckel, O., and Sauberer, F., eds.), Pt. 3, Chap. 4, pp. 122–135. (Sonderheft, *Wetter Leben*).

Dobbins, R. A. (1977). Observations of the barotropic Ekman layer over an urban area. *Boundary Layer Meteorol.* **11,** 39–54.

Druyan, L. M. (1968). A comparison of low-level trajectories in an urban atmosphere. *J. Appl. Meteorol.* **7,** 583–590.

Ekman, V. W. (1905). On the influence of the earth's rotation on ocean currents. *Ark. Mat., Astron. Fys.* **2,** No. 11.

Fontana, P. H., and Bornstein, R. D. (1979). Effect of New York City on the horizontal and vertical structure of sea breeze fronts (Final Report), Vol. I, 53 pp. Dept. of Meteorology, San Jose State Univ., San Jose, California.

Frederick, R. H. (1961). A study of the effect of tree leaves on wind movement. *Mon. Weather Rev.* **89,** 39–44.

Goldreich, Y. (1979a). A possible heat island effect on Bezuidenhout Valley air circulation and temperature distribution. *S. Afr. Geogr. J.* **61,** 123–127.

Goldreich, Y. (1979b). Influence of urbanization on mountain and valley wind regime, *Proc. Sci. Conf. Israel Ecol. Soc., Sede Boker, 10th,* pp. D-19 to D-27.

Hass, W. A., Hoecker, W. H., Pack, D. H., and Angell, J. K. (1967). Analysis of low-level constant volume balloon (tetroon) flights over New York City. *Q. J. Roy. Meteorol. Soc.* **93,** 483–493.

Izumi, Y., and Barad, M. L. (1970). Wind speeds as measured by cup and sonic anemometers and influenced by tower structure. *J. Appl. Meteorol.* **9,** 851–856.

Jackson, P. S. (1978). Wind structure near a city centre. *Boundary Layer Meteorol.* **15,** 323–340.

Jones, P. M., deLarzinaga, M. A. B., and Wilson, C. B. (1971). The urban wind velocity profile. *Atmos. Environ.* **5,** 89–102.

Koschmieder, H. (1935). Der Seewind von Danzig. *Meteorol. Z.* **52,** 491–495.

Kremser, V. (1909). Ergebnisse vieljähriger Windregistrierungen in Berlin. *Meteorol. Z.* **26,** 238–252.

Landsberg, H. (1958). "Physical Climatology," 2nd ed., p. 201. Gray Printing Co., Dubois, Pennsylvania.

Landsberg, H. E. (1979). Atmospheric changes in a growing community (The Columbia, Maryland Experience). *Urban Ecology* **4,** 53–81.

Lee, D. O. (1979). The influence of atmospheric stability and the urban heat island on urban–rural wind speed differences. *Atmos. Environ.* **13,** 1175–1180.

Lettau, H. H. (1969). Note on aerodynamic roughness parameter on the basis of roughness element description. *J. Appl. Meteorol.* **8**, 828–832.

Lettau, H. H. (1970). Physical and meteorological basis for mathematical models of urban diffusion processes. A. C. Stern (ed.), *Symp. Multiple-Source Urban Diffusion Models,* pp. 2-1 to 2-24. Natl. Air Pollut. Control Admin., Washington, D. C.

Loose, T., and Bornstein, R. D. (1977). Observations of mesoscale effects on frontal movement through an urban area. *Mo. Wea. Rev.* **105**, 563–571.

Marsh, K. J. (1969). Measurements of air turbulence in Reading and their relation to Turner's stability categories: *see* Pasquill, F. (1970). Wind structure in the atmospheric boundary layer. *Phil. Trans. Roy. Soc. A* **269**, No. 1199.

Munn, R. E. (1970). Air flow in urban areas. *In* "Urban Climates," *WMO, Tech. Note,* No. 108, pp. 15–39.

Nkedirim, L. C. (1980). Cold air drainage and temperature fields in an urban environment: A case study of topographical influence on climate. *Atmos. Environ.* **14**, 375–381.

Okita, T. (1960). Estimation of direction of air flow from observation of rime ice. *J. Meteorol. Soc. Japan II,* **38**, (4), 207–209.

Patterson, R. D., and Hage, K. D. (1979). Micrometeorological study of an urban valley. *Boundary Layer Meteorol.* **17**, 175–186.

Peschier, Jr., J. (1973). Wind and temperature profiles in an urban area. Rept. No. 33, 32 pp. Atmos. Sci. Gp., Univ. of Texas, Austin.

Rubinshtein, E. C. (1979). Odnorodnost meteorologicheskikh Ryadov vo Vremeni i Prodstanstve v Svya zi c Issledovaniem izmeneniya Klimata. *Gidrometeoizdat (Leningrad),* 79 pp.

Schmauss, A. (1925). Eine Miniaturpolarfront. *Meteorol. Z.* **42**, 196.

Steinhauser, F., Eckel, O., and Sauberer, F. (1959). Klima und Bio Klima von Wien, III Teil. *Welter Leben* **11** (Sonderheft), 135 pp.

Stummer, G. (1939). Klimatische Untersuchungen in Frankfurt am Main und seinen Vororten. *Ber. Meteorol. Geophys. Inst., Univ. Frankfurt,* **5**.

Taylor, R. J. (1952). The dissipation of kinetic energy in the lowest layers of the atmosphere. *Q. J. Roy. Meteorol. Soc.* **78**, 179–185.

Yamamoto, G., and Shimanuki, A. (1964). Profiles of wind and temperature in the lowest 250 metres in Tokyo. *Sci. Rept. Tohoku Univ. Geophys. Ser.* **5** (15), 111–114.

Zanella, G. (1976). Il Clima Urbano di Parma. *Riv. Meteorol. Aeronaut.* **36**, 125–146.

7

Models of Urban Temperature and Wind Fields

There have been a number of approaches to mathematical and numerical simulations of the urban atmospheric boundary layer. There are energy exchange models that are representative of the urban climate and various types of urban land use. Dynamic models are designed to approximate the time-dependent horizontal and vertical wind and temperature fields. They have been variously expanded to indicate distribution and dispersal of pollutants. Although all the proposed schemes offer numerical solutions and are programmed for high-speed computer operations, they all suffer from the fact that initial intraurban conditions and boundary values are almost never known with sufficient density and accuracy to permit more than approximate solutions.

7.1 STATIC MODELS

These models are all based on the energy exchange equation. This was presented previously in Section 4.2. The same symbols are used in this section, although this is somewhat at variance with the preferences of the various authors in this field. For simplicity, a list of symbols is given at the end of this chapter.

Myrup (1969) attempted to represent the urban heat island by a simple energy budget

$$Q_N = Q_E + Q_H + Q_S \tag{7.1}$$

As we have already seen each of these quantities can be either positive or negative. In this equation he substitutes the basic parameters to represent the energy fluxes

$$Q_N = (1 - A)\kappa I_0[\sin \phi \sin \delta + \cos \phi \cos \eta] - Q_L \tag{7.2}$$

$$Q_E = -\rho L K_q \frac{\partial q}{\partial z} \tag{7.3}$$

$$Q_H = -\rho c_p K_H \frac{\partial \Theta}{\partial z} \tag{7.4}$$

$$Q_S = -k_s \frac{\partial T_s}{\partial z} \tag{7.5}$$

Substituting these parameters in Eq. (7 1) represents the essential framework of the model

$$(1 - A)\kappa I_0 |\sin \phi \sin \delta + \cos \phi \cos \eta| - Q_L$$

$$= -\rho L K_q \frac{\partial q}{\partial z} - \rho c_p K_H \frac{\partial \Theta}{\partial z} - k_s \frac{\partial T_s}{\partial z} \tag{7.6}$$

It is obvious that numerical evaluation of the individual parameters is an overwhelming task. It is thus not too surprising that Myrup (1969b) comes to the conclusion that the "simple energy budget approach is not adequate to predict the effect of the human heat load on the temperature of cities." Myrup and Morgan (1972) attempted nonetheless to do this for Sacramento, California, subdivided into 152 squares. They made a detailed urban fabric analysis of the surface characteristics, including land use, population, and traffic. With an elaborate computer program they tried to simulate human skin temperatures and comfort indices in these various environments.

An only slightly different model was proposed by Bach (1970), whose basic energy balance equation was

$$I(1 - A) + Q_{SK} + Q_{L\downarrow} - \epsilon\sigma T_G{}^4 = \pm Q_N = \pm Q_S \pm Q_H \pm Q_E + Q_p \tag{7.7}$$

Bach represents the terms for latent and sensible heat transfer by application of the logarithmic wind law

$$Q_H = \rho c_p k^2 \frac{(u_2 - u_1)(T_1 - T_2)}{(\ln z_2/z_1)^2} \tag{7.8}$$

and

$$Q_E = \rho k^2 \frac{(u_2 - u_1)(q_1 - q_2)}{(\ln z_2/z_1)^2} \tag{7.9}$$

All other energy fluxes are obtained by direct measurement. His conclusions from observations in Cincinnati were that compared with short-wave solar, sky, and the various long-wave fluxes, the anthropogenic flux, including all combustion and metabolic processes, are quite small. In daytime they are about 5 percent of the net radiation and at night about 25 percent.

In an experiment, using infrared emissions measured in midday in an urban area, Outcalt (1972a,b) tested a model based on Myrup's attempt. This simulates a surface equilibrium temperature based on all the fluxes so that

$$Q_N + Q_S + Q_H + Q_E = 0 \tag{7.10}$$

Substituting, as in the Myrup model, the various parameters Outcalt arrives at the surface-energy transfer for the equilibrium condition:

$$(Q_I + Q_{SK})(1 - A) + \epsilon\sigma T_{SK}{}^4 - \epsilon\sigma T_G{}^4 + \frac{\rho k^2 u_2}{[\ln(z_2/z_0)]^2}$$

$$\times \left\{ c(T_2 - \Gamma z_2 - T_G) + L\left[q_2\left(\frac{A_w}{\mathscr{L}}\right)F(T_0) \right] \right\}$$

$$+ \left[\frac{K_S}{(z/2)} \right](T_S - T_G) = 0 \tag{7.11}$$

In the numerical process a subroutine introduces a correction for the departures of the adiabatic exchange coefficient, based on the Richardson number of the form $[1 - a\ Ri]^{1/2}$, where Ri is a nondimensional parameter, characterizing stability:

$$Ri = \frac{g(\partial\Theta/\partial z)}{T_2(\partial u/\partial z)^2}$$

The factor c is an empirically derived constant, set by Outcalt at 32. Comparing calculated surface temperatures with those measured by infrared scanning Outcalt found them verified and concluded that differences in the urban–rural fabric can be produced by land-use variations only.

Very detailed in calculating the effects of radiation and reradiation of urban surfaces is a simulation by Terjung and O'Rourke (1980). The model enabled them to show a not unexpected detailed microstructure of warmer and cooler parts of the urban heat island, in a complex interplay of various radiative and sensible heat exchanges.

7.2 DYNAMIC MODELS OF URBAN FLOW

Quite different is the approach in the dynamic models. In these the existence of the urban heat island is considered as an established fact and the models attempt to simulate the country breeze. Nearly all of them are time-dependent, two-dimensional models, i.e., they represent x, z cross sections over the urban area and surroundings. These models were basically patterned after the earlier developments of simulating secondary circulations, in particular, land and sea breezes. We mention here only one of these prototypes because of its seminal influence. That is the model developed by Estoque (1961, 1963). Based on this work, as expanded by Estoque and Bhumralkar (1970), Delage and Taylor developed a specific two-dimensional urban (1970) flow model in the absence of an external wind field. This places a vertical symmetry plane in the center of the heat island and permits the study of only half the system. The Coriolis parameter was considered unimportant for the dimensions chosen.

A system of four equations then determines the horizontal velocity, the potential temperature, the vertical velocity, and the pressure:

$$\frac{\partial u}{\partial t} = -u \frac{\partial u}{\partial x} - w \frac{\partial u}{\partial z} - \frac{1}{\rho} \frac{\partial p}{\partial x} + \frac{\partial}{\partial z}\left(K_z \frac{\partial u}{\partial z}\right) + K_{XM} \frac{\partial^2 u}{\partial x^2} \quad (7.12)$$

$$\frac{\partial \Theta}{\partial t} = -u \frac{\partial \Theta}{\partial x} - w \frac{\partial \Theta}{\partial z} + \frac{\partial}{\partial z}\left(K_z \frac{\partial \Theta}{\partial z}\right) + K_{XH} \frac{\partial^2}{\partial z^2} \Theta \quad (7.13)$$

$$\rho \frac{\partial u}{\partial x} + \frac{\partial}{\partial z}(\rho w) = 0 \quad (7.14)$$

$$\frac{\partial}{\partial z}\left(\frac{p}{p_0}\right)^{R/c_p} = -\frac{g}{c_p \Theta} \quad (7.15)$$

By integration of (7.15) to the height H, representing the top of the model, the pressure is obtained as

$$p = p_H^{R/c} + \frac{g}{c_p} p_0^{R/c_p} \int_z^H \left(\frac{dz}{\Theta}\right)^{c/R} \quad (7.15a)$$

where p_H is determined by a suitable boundary condition. By definition, the horizontal wind speed at the center of the heat island, $u = 0$. The other boundary conditions are shown in a diagram taken from the paper by Delage and Taylor (1970), Fig. 7.1. Numerical solutions for these equations for a ΔT_{u-r} of 2.5°C, $\partial \Theta / \partial z = 2$°C km^{-1}, $K_z = K_{XH} = 50$ m^2 sec^{-1}, $K_{XM} = 0$ after 90 min running time of the model are shown graphically in Fig. 7.2. An unstable layer with notable vertical wind speed is shown and a well-developed country breeze to the height of about 700 m has developed. At higher levels a compensating return flow is indicated. This result, qualitatively at least not contradicted by observations, is gratifying considering the simplifying assumptions made and the neglect of urban roughness.

Somewhat more elaborate is the model developed by Yu (1973), which includes the Coriolis parameter, a geostrophic wind, and advective motion upwind of the urban boundary. It also uses the energy-budget simulations, essentially as shown previously in Eqs. (7.2)–(7.6), according to Myrup (1969). Other parts of the model use the Navier–Stokes formulations, the continuity condition, the potential temperature distribution, the hydrostatic equation, and the

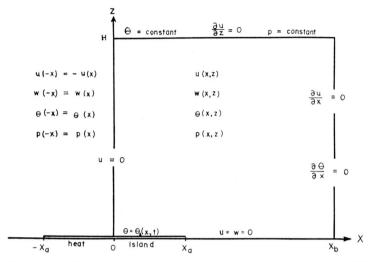

Fig. 7.1 Configurations and boundary conditions of a heat-island circulation model by Delage and Taylor (1970) [reproduced by permission].

turbulent diffusivity for momentum, as shown in Eqs. (7.16)–(7.22):

$$\frac{\partial u}{\partial t} + u\frac{\partial u}{\partial x} + w\frac{\partial u}{\partial z} - fv = -\frac{1}{\rho}\left(\frac{\partial p}{\partial x}\right) + \frac{\partial}{\partial z}\left(K_M\frac{\partial u}{\partial z}\right) \qquad (7.16)$$

$$\frac{\partial v}{\partial t} + u\frac{\partial w}{\partial x} + w\frac{\partial v}{\partial z} + fu = -\frac{1}{\rho}\left(\frac{\partial p}{\partial y}\right) + \frac{\partial}{\partial z}\left(K_M\frac{\partial v}{\partial z}\right) \qquad (7.17)$$

$$\frac{\partial u}{\partial x} + \frac{\partial w}{\partial z} = 0 \qquad (7.18)$$

$$\frac{\partial \Theta}{\partial t} + u\frac{\partial \Theta}{\partial x} + w\frac{\partial \Theta}{\partial z} = \frac{\partial}{\partial z}\left(K_H\frac{\partial \Theta}{\partial z}\right) \qquad (7.19)$$

$$\frac{\partial T_S}{\partial t} = \frac{\partial}{\partial z}\left(K_S\frac{\partial T_S}{\partial z}\right) \qquad (7.20)$$

$$\frac{\partial}{\partial z}\left(\frac{p}{p_0}\right)^{R/c_p} = -\frac{g}{c_p\Theta} \qquad (p_0 = 1000\ \text{mb}) \quad (7.21)$$

$$K_M = K_H \begin{cases} l^2\left[\left(\frac{\partial u}{\partial z}\right)^2 + \left(\frac{\partial v}{\partial z}\right)^2\right]^{1/2}(1 + \beta_0\ \text{Ri}), & \text{for}\quad \text{Ri} < 0 \\[2ex] l^2\left[\left(\frac{\partial u}{\partial z}\right)^2 + \left(\frac{\partial v}{\partial z}\right)^2\right]^{1/2}(1 - \beta_0\ \text{Ri})^{-1}, & \text{for}\quad \text{Ri} > 0 \end{cases} \qquad (7.22)$$

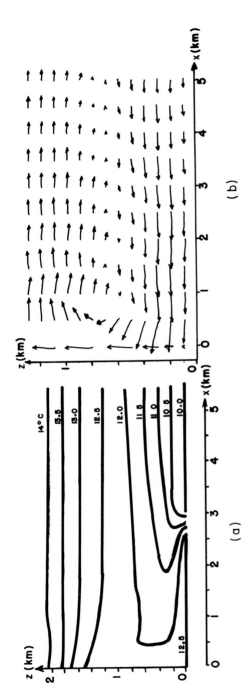

Fig. 7.2 Half-temperature cross section over heat island (a), zero on x axis in center of heat island and wind field (b), after 90 min of model time (from Delage and Taylor, 1970).

In Eq. (7.17),

$$-\frac{1}{\rho}\left(\frac{\partial p}{\partial y}\right) = u_{Gx}$$

i.e., geostrophic wind component in the x direction. In Eq. (7.22),

$$l = \frac{k(z + z_0)}{1 + [k(z + z_0)/\lambda]}, \qquad \text{mixing length}$$

$$\lambda = 0.00027 u_G f^{-1}$$

$$\beta_0 = -3$$

It should be noted here that by using a representation of the turbulent diffusivity (7.22) in the form given by Estoque and Bhumralkar (1970), Yu introduces the roughness coefficient z_0 into the model. The assumption that the turbulent diffusivities for heat and momentum are equal simplifies the system of equations, but in nature these parameters may have different values. The boundary conditions for the model chosen by Yu are at the upwind boundary: the vertical distributions of the horizontal wind components and of the potential temperature. At the soil–air interface all wind components are zero. In the soil, the lower boundary is set at -50 cm and the distribution of temperature in the x direction assumed. The upper boundary is fixed at 1000 m, the wind speed in the x direction is assumed to be the geostrophic wind, the vertical wind speed and the wind component in y direction are set at zero. The potential temperature and pressure are known. The assumption that the upper boundary acts as a barrier for momentum and heat flux is not always fulfilled, and the complete exclusion of moisture exchange, at least at the soil interface, may introduce small errors.

After transformation into finite difference form, Yu calculates the changes taking place in the system. One of his case studies starts with the following conditions. At 1000 m: $u_G = 10$ m sec^{-1}, $\Theta = 312$ K, and $p = 900$ mb. Further, at 300 m, Θ is specified at 310 K, with a ground inversion and 307.4 K at the surface. The soil temperature at the lower boundary (-50 cm) in the soil was chosen at 295 K. For a nocturnally developing heat island, the perturbation potential temperature and vertical velocities were calculated. The roughness parameter z_0 was gradually increased from 5 cm (grid unit 1 along the x direction) in the rural area to 50 cm in the upwind suburban area (grid unit 5), to 150 cm in the urban center (grid unit 9), tapering off again downwind to 5 cm (grid unit 17). Figure 7.3 shows the patterns as they appear after 5 hr of operation. A 3.6°C temperature in the city center at the surface is shown; at the usual 2-m meteorological shelter height a temperature surplus of 1°C was calculated. At 1 m it was 1.5°C and at 300 m the excess vanished. Downwind there was a slight temperature decrease indicated at about 200 m. The horizontal wind speed in the city decreased at a 10-m height by 35 percent of the upwind values quite in accordance with observations. At 300 m the decrease vanished. The vertical velocity perturbation at about 200 m, somewhat upwind of the city center, was probably correct in the order of magnitude for a nocturnal situation, where the mechanical friction effect rather than thermal instability was the cause for the perturbation (Wagner and Yu, 1972).

Very similar models were developed by Atwater (1975) and Gutman and Torrance (1975). Of general interest are the general model parameters used by these authors. They are shown in Table 7.1. It should be noted that most of these values are in accordance with measurements, but the restriction to categories such as rural, suburban, and urban introduces a simplification that restricts the models to a gross generalization of reality in large cities. There building density, building heights, parks, lakes, rivers, and other diversities are far more complex than these simple models can handle. The parameter values adopted by various authors also diverge, as Table 7.1 shows. Some of these values, as well as the meteorological boundary conditions, are fairly arbitrary and hence some of the conclusions drawn from the model results are at variance with observed facts. An example is the statement by Gutman and Torrance that: "Without heat addition, evaporation and substrate material create a negligible heat island." Even under the most limited conditions this cannot be verified by the facts.

The most elaborate and, generally, most realistic model is that developed by Bornstein (1972, 1975). He labels his model URBMET and uses the vorticity form of the boundary-layer equations, as follows:

$$\frac{\partial \xi}{\partial t} + \frac{\partial (u\xi)}{\partial x} + \frac{\partial (w\zeta)}{\partial z} = -\frac{g}{T_m}\frac{\partial \Theta'}{\partial K} + f\frac{\partial v}{\partial z} + \frac{\partial^2}{\partial x^2}(K_M\zeta) \quad (7.23)$$

where the primed quantity represents the perturbation from the undisturbed state and the subscript m stands for the space average.

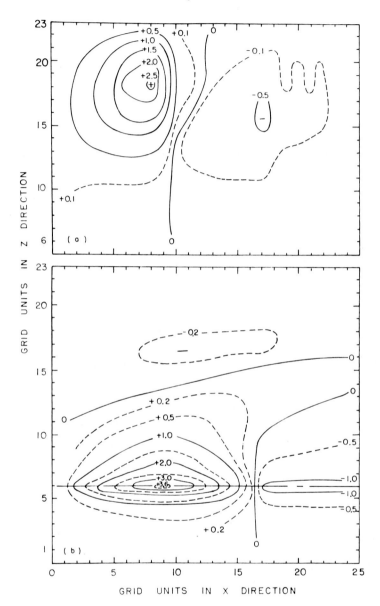

Fig. 7.3 Distribution of vertical velocity (cm sec^{-1}) (a) and perturbation potential temperature (°C) (b) after 5 hr of model time of urban heat-island circulation. Urban center is located at grid point 9 (from Yu, 1973).

TABLE 7.1

Some of the Parameter Values Used in Urban Models[a]

Element	Rural	Suburban	Urban
Coriolis parameter (sec^{-1})	0.00001	0.00001	0.00001
Surface thermal diffusivity (m^2 sec^{-1})	0.5×10^{-6}	1.25×10^{-6} (1.0×10^{-6})	2.0×10^{-6} (1.5×10^{-6})
Surface density (kg m^{-3})	2000	2000	2000
Surface heat capacity (kJ m^3 K^{-1})	1088.5	1120.0	1151.4
Roughness parameter (m)	0.01	0.5 (0.25)	1.0 (3.0)
Moisture parameter	0.9	0.55	0.2 (0.8)
Albedo	0.2	0.2	0.2

[a] Where the sources show different values, the Gutman–Torrance values are shown in parentheses. Not all parameters are given in both studies. (After Atwater, 1975; Gutman and Torrance, 1975.)

Using the stream function the following definitions apply:

$$u = \frac{\partial \psi}{\partial z}, \qquad w = -\frac{\partial \psi}{\partial x}, \qquad \zeta = \frac{\partial^2 \psi}{\partial z^2}$$

In addition to the Richardson number, Bornstein also employs the Monin–Obukhov length for stability consideration, given by

$$\mathscr{L} = \frac{u_* \Theta_m}{g k \Theta_*}$$

Also introduced are vertical variations of the turbulent transfer coefficients for heat and momentum:

$$\left(\frac{\partial K_M}{\partial z}\right)_h = \frac{K_M(h)}{(h + z_0)} \left[1 - \frac{h + z_0}{\phi} \left(\frac{\partial \phi}{\partial z}\right)_h \right] \qquad (7.24)$$

and

$$\left(\frac{\partial K_H}{\partial z}\right)_h = \frac{K_H(h)}{(h + z_0)} \left[1 - \frac{2(h + z_0)}{\phi} \left(\frac{\partial \phi}{\partial z}\right)_h \right] \qquad (7.25)$$

where Φ is a stability parameter

$$\phi = 1 + a\left(\frac{z + z_0}{\mathscr{L}}\right)$$

for forced convection, and, for free convection,

$$\Phi = 1\left(\frac{c}{3}\right)^{1/2}\left|\frac{z + z_0}{\mathscr{L}}\right|^{-1/6}$$

For the model calculations, the height of the constant flux layer is set at 25 m, the height H_{*1} where the turbulent transfer coefficients become small and constant, 1050 m, and H the top of the transition layer is set at 1900 m. For the constant flux layer the model uses a set of 13 equations. These are shown in Table 7.2, side by side, for two stability conditions.

Bornstein carried through a number of sample calculations for nocturnal conditions, with rather weak geostrophic winds. He used 0.5 m as a rural roughness parameter, and for the urban area a value of 3 m was used. Temperature change in the city for the first 6 hr of simulation was assumed to be zero, while in the rural area an hourly drop from 1800 to 2400 hr of 1°C was postulated. The results for the departures of horizontal and vertical wind fields are shown in Fig. 7.4a,b. The vertical wind perturbations are an order of magnitude smaller than the horizontal values. The latter show weakening at the windward edge and in the lowest layer above the urban and leeward area. Upwind at 400 m there is a 10 percent wind reduction and downwind at 300 m there is a 10 percent increase. The Coriolis terms were very small in the $\sim$90-km cross section and their neglect in earlier models was quite justifiable. Mechanical and thermal turbulence are the factors affecting the wind field most.

A further elaboration of the URBMET model was carried out by Bornstein and Robock (1976) for nocturnal conditions. They introduced a variable, but equal, time step for the advective and diffusive processes. They also varied the turbulent diffusivities both in space and time. These devices enabled them to cut computer time in half without appreciable changes in the simulated values, compared with use of constant time steps. Also introduced into this amended URBMET model was an anthropogenic moisture source over the urban heat island. This enhanced the buoyancy of the air by $\sim$10 percent in the lowest few hundred meters.

Vukovich (1971, 1973) also developed a two-dimensional model without the roughness parameter but encompassing the diurnal variation of the heating function. He and his collaborators (Vukovich *et al.*, 1976) transformed this into a three-dimensional model, which they applied to St. Louis conditions.

They used the usual meteorological equations of motion and con-

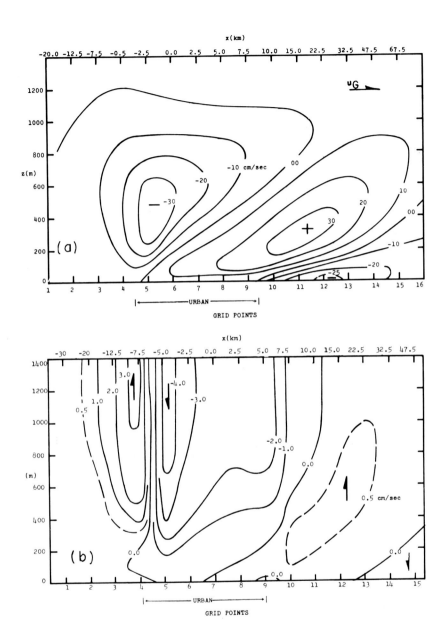

Fig. 7.4 Departures of horizontal (a) and vertical wind field (b) (cm sec^{-1}) from an undisturbed flow during nocturnal conditions over an urban heat island, using Bornstein's URBMET model (from Bornstein and Robock, 1976).

TABLE 7.2

Constant Flux Layer Equations for URBMET Model

For stable case Ri ≥ 0	For lapse, free convection, Ri < 0
(1) $\quad r = 1$	$r = (c/3)^{1/2} \left\lvert \dfrac{z + z_0}{\mathscr{L}} \right\rvert^{1/6}$
(2) $\quad \phi = 1 + a[(z + z_0)/\mathscr{L}]$	$\phi = (c/3)^{1/2} \left\lvert \dfrac{z + z_0}{\mathscr{L}} \right\rvert^{-1/6}$
(3) $\quad \dfrac{\partial u}{\partial z} = \dfrac{u_*}{k(z + z_0)} \left[1 + a\left(\dfrac{z + z_0}{\mathscr{L}} \right) \right]$	$\dfrac{\partial u}{\partial z} = \dfrac{u_*}{k(z + z_0)} (c/3)^{1/2} \left\lvert \dfrac{z + z_0}{\mathscr{L}} \right\rvert^{-1/6}$
(4) $\quad \dfrac{\partial \Theta^\circ}{\partial z} = \dfrac{\Theta^*}{k(z + z_0)} \left[1 + a\left(\dfrac{z + z_0}{\mathscr{L}} \right) \right]$	$\dfrac{\partial \Theta'}{\partial z} = \dfrac{\Theta_*}{k(z + z_0)} 3/c \left\lvert \dfrac{z + z_0}{\mathscr{L}} \right\rvert^{-1/6}$
(5) $\quad u(z) = \dfrac{u_*}{k} \left(\ln \dfrac{z + z_0}{z_0} + a \dfrac{z}{\mathscr{L}} \right)$	$u(z) = \dfrac{6u_*}{k} (c/3)^{1/2} \left(\left\lvert \dfrac{z_0}{\mathscr{L}} \right\rvert^{1/6} - \left\lvert \dfrac{z + z_0}{\mathscr{L}} \right\rvert^{-1/6} \right)$
(6) $\quad \Theta(z) = \Theta'(0) + \dfrac{\Theta^*}{k} \left(\ln \dfrac{z + z_0}{z_0} + a \dfrac{z}{\mathscr{L}} \right)$	$\Theta(z) = \Theta'(0) + \dfrac{c\Theta_*}{k} \left(\left\lvert \dfrac{z_0}{\mathscr{L}} \right\rvert^{-1/3} - \left\lvert \dfrac{z + z_0}{\mathscr{L}} \right\rvert^{-1/6} \right)$

$$(7) \quad u_* = \frac{k(u+z_0)[1 - a\text{Ri}(3h/4)u]3h/2}{\left(1 + a\dfrac{h+z_0}{\mathcal{L}}\right)\left(\dfrac{h}{z_0+h}\right) + \ln\left(\dfrac{0.5h+z_0}{z_0}\right) + \dfrac{0.5ah}{\mathcal{L}}}$$

$$u_* = \frac{k(3/c)^{1/2}u(3h/2)}{6\left(\left|\dfrac{z_0}{\mathcal{L}}\right|^{-1/6} - \left|\dfrac{0.5h+z_0}{\mathcal{L}}\right|^{-1/6} + \left(\dfrac{h}{h+z_0}\right)\left|\dfrac{z+z_0}{\mathcal{L}}\right|^{-1/3}\right)}$$

$$(8) \quad \Theta_* = \frac{k(h+z_0)[1 - a\,\text{Ri}(3h/4)][\Theta'(3h/2) - \Theta(0)]}{k + (h+z_0)\ln\left(\dfrac{0.5h+z_0}{z_0}\right)}$$

$$\Theta_* = \frac{3(k/c)[\Theta'(3h/2) - \Theta'(\sigma)]}{3\left(\left|\dfrac{z_0}{\mathcal{L}}\right|^{-1/3} - \left|\dfrac{0.5h+z_0}{\mathcal{L}}\right|^{-1/3} + \left(\dfrac{h}{h+z_0}\right)\left|\dfrac{h+z_0}{\mathcal{L}}\right|^{-1/3}\right)}$$

$$(9) \quad K_M(z) = [k(z+z_0)]^2|\partial u/\partial z|[1 - a\,\text{Ri}(3h/4)]^2$$

$$K_M(z) = (3/c)[k(z+z_0)]^2|\text{Ri}(3h/4)|^{1/3}|\partial u/\partial z|$$

$$(10) \quad K_H(z) = [k(z+z_0)]^2|\partial u/\partial z|[1 - a\,\text{Ri}(3h/4)]^2$$

$$K_H(z) = (3/c)^{3/2}[k(z+z_0)]^2|\text{Ri}(3h/4)|^{1/2}|\partial u/\partial z|$$

$$(11) \quad (\partial\phi/\partial z)_h = a\mathcal{L}$$

$$\left(\frac{\partial\Phi}{\partial z}\right)_h = -\frac{\Phi}{6(h+z_0)}$$

$$(12) \quad \left(\frac{\partial K_M}{\partial z}\right)_h = \frac{K_M(h)}{(h+z_0)}[1 - a\,\text{Ri}(h)]$$

$$\left(\frac{\partial K_M}{\partial z}\right)_h = \frac{7K_M(h)}{6(h+z_0)}$$

$$(13) \quad \left(\frac{\partial K_H}{\partial z}\right)_h = \frac{K_H(h)}{(h+z_0)}[1 - a\,\text{Ri}(h)]$$

$$\left(\frac{\partial K_H}{\partial z}\right)_h = \frac{4K_H(h)}{3(h+z_0)}$$

TABLE 7.3

Influence on Wind Divergence of Heat Island, Roughness, and Horizontal Diffusion[a]

Heat-Island intensity $(\Delta T_{u-r} \,^\circ C)$	Divergence $(sec^{-1} \times 10^{-3})$	Roughness z_0 (m)	Divergence $(sec^{-1} \times 10^{-4})$	Diffusion coeff. at center $(m^2 \, sec^{-1})$	Divergence $(sec^{-1} \times 10^{-4})$
1.5	−0.2	0.5	5.5	900	8.0
3.0	−0.6	1.0	6.0	1200	6.0
4.5	−1.4	2.0	6.5	1600	5.0
6.0	−2.0				

[a] After Vakovich and Dunn (1978). (Maximum divergence at 100 m.)

servation as used for general meteorological predictions. Roughness and turbulent diffusion coefficients are included as in the two-dimensional models. Using a 72×72 km grid space and observed initial conditions, they calculated the wind convergence, updraft, and heat plume configuration. A rather detailed surface wind regime over the area could thus be simulated, with the ultimate aim to predict dispersal of pollutants.

In a sensitivity analysis based on this model Vukovich and Dunn (1978) studied the effect of various parameters on the wind divergence. Their principal results are shown in Table 7.3.

The model clearly demonstrated that heat-island intensity and vertical stability had the most pronounced influence on the urban circulation. A somewhat simplified three-dimensional model was developed by Uliasz (1979). A numerical calculation for two relatively weak geostrophic wind speeds yielded the wind perturbation at several levels above roof height and lapse rates. Wind-speed reductions and accelerations were in agreement with other modeling results and fairly representative of observations.

7.3 MODEL APPLICATIONS TO AIR POLLUTION STUDIES

We have already seen in Chapter 3 that very simple models are quite capable of estimating, to a first approximation, the pollution concentration in an urban area (Hanna, 1978). The easy formulation is

$$\chi = C \frac{q_\mathrm{p}}{u}$$

which simply states that pollutant concentration is proportional to emission strength and inversely proportional to the wind speed, multiplied by a constant. This constant C has been empirically set at 50. Apparently, one can improve on the predictive value of this simple scheme by introducing values for different seasons. This is obviously related to the different seasonal stability conditions. For Pretoria, South Africa, Zib (1980) used a winter value of 55 and a summer value of 20. His measured and predicted values of SO_2

showed a correlation coefficient of 0.7 in winter (explaining 49 percent of the variance) and 0.85 in summer (explaining 72 percent of the variance).

A somewhat more elaborate, but still simple model was proposed by Leahey (1972):

$$\chi(h_m) = \frac{\int_0^x q_p \, dx}{u h_m} \tag{7.26}$$

which integrates the emissions along an air-parcel trajectory and calculates the pollutant concentration in the air column up to the mixing height. Morning concentrations of SO_2 in New York City predicted by the formula showed a correlation coefficient of 0.83, explaining about 69 percent of the variance.

Clearly, in all urban air pollution problems the mixing height h_m is a critical element. Several advective models, closely resembling each other have been developed and used by Summer (1965), Kalma (1974), and Henderson-Sellers (1980). Neglecting friction and starting with a steady-state energy equation of a column of stable, rural air moving to an urban area:

$$c_p \rho u \left[\int_0^{h_m} \left(T_G + \frac{\partial \Theta}{\partial z} h_m - \Gamma_z \right) dz + \int_{h_n}^{h_z} (T_G - \gamma z) \, dz \right]$$
$$= \int \int Q_s n \, dA_C \tag{7.27}$$

where the potential temperature gradient refers to the stable rural air. This can be integrated to yield an expression for the mixing height (interchangeably designated as "mixing depth"):

$$h_m^2(x) = \frac{2}{c_p \rho u (\partial \Theta / \partial z)} \int_0^x Q_S \, dx \tag{7.28}$$

This applies to flat terrain but Henderson-Sellers (1980) developed it further for irregular terrain, $h_m + z' = h_m'$ where the factor z' is the level of the ground below an arbitrary level $z = 0$, and positive z values above this level, so that

$$T_G (z = 0) - \Gamma h_m' = (T_G - \Gamma h_m)[T_G + \Gamma z'/T_G]^{[-8/(\partial \Theta/\partial z)]} \tag{7.29}$$

For neutral stability the mixing height in irregular terrain reduces to

$$h_m' = (T_G - h)[lT_G + z^1)/T_G]^{[-8/(\partial \Theta/\partial z)]} - T_G \tag{7.30}$$

Elaborate attempts have been made to model pollution concentrations in urban areas by including time-dependent pollutant concentrations and radiative energy interactions. Such work has been presented by Viskanta et al. (1976) and Viskanta and Weirich (1979). Theirs is essentially a primitive equation model based on the conservation of mass, momentum, energy, and pollution species. To this they add the radiative interactions with urban pollutants. The boundary conditions are adjusted for urban–rural and seasonal differences. Turbulent diffusivity and roughness adjustments are included but not as elaborately as in the Bornstein URBMET model. The basic set of equations is as follows:

mass conservation

$$\frac{\partial(\rho u)}{\partial x} + \frac{\partial(\rho w)}{\partial z} = 0 \tag{7.31}$$

momentum conservation in x, y, z directions

$$\left(\frac{\partial u}{\partial t} + u\frac{\partial u}{\partial x} + w\frac{\partial u}{\partial z}\right) = f(v - v_G) + \frac{\partial}{\partial z}\left(K_{M_z}\frac{\partial u}{\partial z}\right) \tag{7.32}$$

$$\left(\frac{\partial v}{\partial t} + u\frac{\partial v}{\partial x} + w\frac{\partial v}{\partial z}\right) = f(u - u_G) + \frac{\partial}{\partial z}\left(K_{M_z}\frac{\partial v}{\partial z}\right) \tag{7.33}$$

$$\frac{\partial p}{\partial z} + \rho g = 0 \tag{7.34}$$

energy

$$\frac{\partial \Theta}{\partial t} + u\frac{\partial \Theta}{\partial x} + w\frac{\partial \Theta}{\partial z} = \frac{\partial}{\partial z}\left(K_{H_z}\frac{\partial \Theta}{\partial z}\right)$$

$$- \left[\frac{\partial Q_N}{\partial z} - Q_P\right]\left(\frac{P_0}{p}\right)^{R/c_p} \Big/ \phi c_p \tag{7.35}$$

pollutant species

$$\frac{\partial \chi_n}{\partial t} + u\frac{\partial \chi_n}{\partial x} + w\frac{\partial \chi_n}{\partial z} = \frac{\partial}{\partial z}\left(K_{z\chi_n}\frac{\partial \chi_n}{\partial z}\right) + \sum \chi_n \tag{7.36}$$

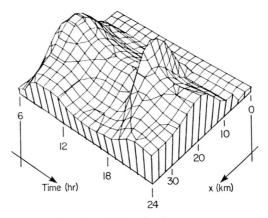

Fig. 7.5   Calculated diurnal variation of pollutant concentration at 2-m height in an urban area during summer (from Viskanta and Weirich, 1979).

soil energy equation

$$\frac{\partial T_S}{\partial t} = 1 - A_S\left(\frac{\partial^2 T_S}{\partial z^2}\right) \tag{7.37}$$

The rather complex radiative effects of pollutants on the various radiative fluxes are not reproduced here. They can be found in Bergstrom and Viskanta (1974). Compared with the other fluxes these are very secondary influences on pollution concentrations. The rather elaborate simulations of pollutant concentrations from this model offer no particular surprises. A case of the diurnal variation in summer of calculated pollutant concentration in an urban area is shown in Fig. 7.5. The diurnal variation is well marked with high nocturnal values and lower concentrations in midday. The turbulent diffusivities, based on the lapse rates, are the major controlling factors.

Even these detailed models do not yet incorporate the many chemical interactions and chemical changes that take place in urban atmospheres. This and the inadequacy of information on boundary conditions make it rather doubtful that the use of such intricate and complex numerical simulations is justified for any practical application. Moreover, their predictive value is very limited in view of the uncertainties in the basic meteorological variables.

List of Symbols

a	Constant in free convection profiles ($=0.3$)	q	Specific humidity
A	Albedo	q_p	Pollutant flux
A_C	City area	Q_E	Heat flux of condensation or evaporation
A_S	Albedo of soil	Q_H	Sensible heat flux
A_W	Area of wet surface	Q_I	Short-wave solar radiation at surface
c	Constant in forced convection profiles	Q_L	Net long-wave radiative flux at surface
c_p	Specific heat at constant pressure	Q_N	Total net heat flux
C	Constant in air pollution concentration estimates	Q_S	Heat flux into or out of surface
		Q_{SK}	Short-wave sky radiation
f	Coriolis parameter	R	Gas constant
g	Acceleration of gravity	Ri	Richardson number
h	Height of constant flux layer	t	Time
h_m	Height of mixing layer	T	Air temperature
h_m^1	Height of mixing layer for ground below an arbitrary level	T_G	Ground surface temperature
		T_S	Soil temperature (at $\frac{3}{2}$, z is depth of penetration of diurnal variation)
h_s	Height of stable surface—near layer	u	Horizontal wind speed in x direction
H	Height of layer or model	u_G	Geostrophic wind speed
I_0	Solar energy received at upper boundary of atmosphere ("solar constant")	u_*	Friction velocity
		v	Horizontal wind component in y direction
k	von Kármán constant (~0.4)	w	Vertical wind speed
K	Turbulent diffusion coefficient	x	Coordinate and distance in direction of u
K_E	Turbulent diffusivity for water vapor	z	Vertical distance from surface
K_H	Turbulent diffusivity for heat	z_0	Roughness parameter
K_M	Turbulent diffusivity for momentum	Z	Depth in soil where diurnal temperature variation vanishes
K_S	Diffusivity for heat in soil		
K_X	Horizontal turbulent transfer coefficient	β	Lapse rate in stable air mass
K_z	Vertical turbulent transfer coefficient	Γ	Lapse rate in mixed layer
		δ	Solar declination
l	Prandtl mixing length	ϵ	Emissivity
L	Latent heat of vaporization	ζ	Vorticity
$\mathscr{L}$	Monin–Obukhov length	η	Hour angle of sun
n	Unit area	Θ	Potential temperature
p	Pressure	Θ^1	Potential temperature perturbation
p_0	Surface pressure		

Θ_* Friction potential temperature χ Pollutant concentration
κ Atmospheric transmission χ_{h_M} Pollutant concentration in
   coefficient for short waves   mixed air column
ρ Air density χ_n Concentration of pollutant n
λ Latitude $\Sigma\chi_n$ Volumetric production of
ϕ Stability parameters   pollutant species n
Φ Characteristic eddy length ψ Stream function
Subscripts 1 and 2 refer to element values at two levels

References

Atwater, M. A. (1975). Thermal changes induced by urbanization and pollution. *J. Appl. Meteorol.* **14**, 1061–1071.

Bach, W. (1970). An urban circulation model. *Arch. Met. Geophys. Biokl. Ser. B* **18**, 155–168.

Bergstrom, R. W., and Viskanta (1974). Spherical harmonics approximation for radiative transfer in pollutant atmospheres. *In* "Thermophysics and Spacecraft Thermal Control" (R. C. Herring, ed.), pp. 23–40. MIT Press, Cambridge.

Bornstein, R. D. (1972). Two dimensional non-steady numerical simulations of night-time flow of a planetary boundary layer over a rough warm city. *Conf. Urban Environ.; Conf. Biometeorol., Philadelphia, 2nd.,* pp. 89–94. Preprints, Am. Meteorol. Soc., Boston.

Bornstein, R. D. (1975). The two-dimensional URBMET boundary layer model. *J. Appl. Meteorol.* **14**, 1459–1477.

Bornstein, R. D., and Robock, A. D. (1976). Effects of variable and unequal time steps for advective and diffusive processes in simulations of the urban boundary layer. *Mon. Weather Rev.* **104**, 260–267.

Delage, Y., and Taylor, P. A. (1970). Numerical studies of heat island circulations. *Boundary Layer Meteorol.* **1**, 201–226.

Estoque, M. A. (1961). A theoretical investigation of the sea breeze. *Q. J. Roy. Meteorol. Soc.* **87**, 136–146.

Estoque, M. A. (1963). A numerical model of the atmospheric boundary layer. *J. Geophys. Res.* **68**, 1103–1113.

Estoque, M. A., and Bhumralkar, C. M. (1970). A method for solving the planetary boundary layer equations. *Boundary Layer Meteorol.* **1**, 169–194.

Gutman, D. P., and Torrance, K. E. (1975). Response of the urban boundary layer to heat addition and surface roughness. *Boundary Layer Meteorol.* **9**, 217–233.

Hanna, S. R. (1978). Urban modelling of inert substance. *In* "Air Quality Meteorology and Atmospheric Ozone" (A. L. Morris and R. Barras, eds.), pp. 262–275. Am. Soc. for Testing Materials, Philadelphia.

Henderson-Sellers, A. (1980). A simple numerical simulation of urban mixing depths. *J. Appl. Meteorol.* **19**, 215–218.

Kalma, J. D. (1974). An advective boundary-layer model applied to Sydney, Australia. *Boundary Layer Meteorol.* **6**, 351–361.

Leahey, D. M. (1972). An advective model for predicting air pollution within an urban heat island with applications to New York City. *J. Air Pollut. Control Assoc.* **22,** 548–550.

Myrup, L. O. (1969). A numerical model of the urban heat island. *J. Appl. Meteorol.* **8,** 908–918, Corrigendum, *ibid.* **9,** (1970), p. 54.

Myrup, L. O., and Morgan, D. L. (1972). Numerical model of the urban interface. Vol. I: the City-Surface Interface. *Atm. Sci.,* No. 4, 237 pp. Univ. of California, Davis.

Outcalt, S. I. (1972a). The development and application of a simple digital surface climate simulator. *J. Appl. Meteorol.* **11,** 629–636.

Outcalt, S. I. (1972b). A reconnaissance experiment in mapping and modeling the effect of land use on urban thermal regimes. *J. Appl. Meteorol.* **11,** 1369–1373.

Summer, P. W. (1965). An urban heat island model, its role in air pollution problems with applications to Montreal. Paper presented at *Can. Conf. Micrometeorol., 1st, Toronto, 12–14 April.*

Terjung, W. H., and O'Rourke, P. A. (1980). Simulating the causal elements of urban heat islands. *Boundary Layer Meteorol.* **19,** 93–118.

Uliasz, M. (1979). Wply miasta na dynamike granicznej warstwy atmosfery (Urban effect on the dynamics of the atmospheric boundary layer). *Przegl. Geofiz.* **24** (32), 57–64.

Viskanta, R., Johnson, R. O., and Bergstrom, R. W., Jr. (1976). Modelling of temperature and pollutant concentration distributions in urban atmospheres. *Trans. ASME, J. Heat Transfer,* Nov. 1976, 662–669.

Viskanta, R., and Weirich, T. L. (1979). Effects of pollutants and urban parameters on atmospheric dispersion and temperature. *U. S. Environ. Prot. Agency, Environ. Sci. Res. Lab.,* EPA-600/4-79-012, 127 pp. Research Triangle Park, North Carolina 27711.

Vukovich, F. M. (1971). Theoretical analysis of mean wind and stability on heat island circulation of an urban complex. *Mon. Weather Rev.* **99,** 919–926.

Vukovich, F. M. (1973). A study of the atmospheric response due to a diurnal function characteristic of an urban complex. *Mon. Weather Rev.* **101,** 467–474.

Vukovich, F. M., and Dunn, J. W. (1978). A theoretical study of the St. Louis heat island: Some parameter variations. *J. Appl. Meteorol.* **17,** 1585–1599.

Vukovich, F. M., Dunn, J. W., and Crissman, B. (1976); A theoretical study of the St. Louis heat island: The wind and temperature distribution. *J. Appl. Meteorol.* **15,** 417–440.

Wagner, N. K., and Yu, T-W. (1972). Heat island formation: a numerical experiment. *Conf. Urban Environ.; Conf. Biometeorol., Philadelphia, 2nd.,* pp. 83–87. Preprints, Am. Meteorol. Soc., Boston.

Yu, T-W. (1973). Two-dimensional time-dependent numerical simulation of atmospheric flow over an urban area. Rept. No. 32, 114 pp. Atmos. Science Group, Univ. of Texas, Austin, Texas.

Zib, P. (1980). Seasonal variability of the simple urban dispersion model. *J. Air Pollut. Control Assoc.* **30,** 35–37.

8

Moisture, Clouds, and Hydrometeors

The temperature and wind fields of urban areas are fairly well understood and well documented by observations. Physical models are available to support the empirical facts. In contrast, the urban moisture field is controversial in many respects. Among the only partially answered questions are the following:

(1)   Do human activities contribute to the atmospheric moisture balance in urban areas?

(2)   Do metropolitan areas change precipitation patterns—and in which direction?

(3)   Are such changes as may exist chemical, microphysical, thermodynamical in origin, or are they a mixture of these?

Even the well-designed METROMEX project yielded only limited answers, partly because it covered only the summer season. Some of the observed facts are essentially only consequences of the urban heat island, but other uncertainties need further investigation.

8.1 THE URBAN HUMIDITY FIELD

Several factors affect the urban humidity field. One of these is the radical change in surface. Impermeable areas abound: Roofs and pavements. These are designed to carry all precipitation off as rapidly as possible and convert it into runoff. In inner cities there is little, if any, vegetation left and thus evapotranspiration is radically reduced. On the other hand, there are many combustion processes, especially those using hydrocarbons that will have water vapor as one of the end products. In addition, many industrial activities release steam and evaporate vast amounts of cooling water. Power plants are a prime example. Their water-vapor release is generally at a higher level from cooling towers, whereas much of the water vapor from hydrocarbons emanates from motor vehicle exhaust essentially at ground level. No adequate inventories of these various vapor effluents exist.

The preponderance of evidence shows that the relative humidity in urban areas at the usual 2-m shelter level is lower in the city than in the country. Paired observations at airports in the city confines and in the country show this clearly. Table 8.1 shows this for New York and Chicago, where airports at the same sea-level elevation clearly show the effect.

In the low-humidity classes the in-town airport has more cases than the fringe field. In contrast, the outlying airport has more cases of high-humidity values than the urban airport. This becomes even more evident if one looks at the diurnal variation of the very high humidities, as shown in Table 8.2.

The overwhelmingly larger number of higher values of high humid-

TABLE 8.1

Frequency of Relative Humidities (in percent) for Various Intervals at Urban and Rural Airports

Locality	0–29	30–49	50–69	70–79	80–89	90–100
La Guardia, New York	2	22	37	14	14	11
Kennedy, New York	1	16	33	16	16	18
Midway, Chicago	2	14	34	20	18	10
O'Hare, Chicago	1	11	32	22	20	15

TABLE 8.2

Annual Frequency[a] **at Specific Hours of Humidities between 90 and 100 Percent Relative Humidities at Urban and Rural Airports**

Locality	Hour of day					
	00	04	08	12	16	20
La Guardia, New York	14	18	11	7	7	10
Kennedy, New York	24	30	17	10	10	18
Midway, Chicago	12	18	10	5	5	8
O'Hare, Chicago	24	27	13	7	7	11

[a] In percent of all observations.

ities at the outlying airport compared to the urban airfield is principally a reflection of the heat island. Zanella (1976) made a comparison for Parma, a city of 170,000 inhabitants for a 25-yr period. The average heat island there is 1.4°C and the average reduction of relative humidity between city and airport is 5 percent, with a maximum of 7 percent in January and 3 percent in the months of July–September.

Landsberg and Maisel (1972) found in daytime a 4 percent humidity reduction and deduced that half of this value is simply a reflection of higher temperatures of the heat island and the other half is attributable to reduced evapotranspiration. Kratzer (1956) reported for Munich a slightly increased vapor pressure in the city for winter but a reduction of about -0.8 mbar in summer compared to outlying districts. The summer results for METROMEX yielded mixing ratios for various stations at surface level. These showed also a lower value of 0.5 g kg^{-1}. Hilberg (1978) reported these for central St. Louis and for a farm station, east of the city, for three classes of cloudiness. His results are shown in Fig. 8.1. The city and the country show a fairly even course throughout the diurnal cycle on cloudy days. On the clear days, there is not only a well-marked diurnal variation but also a clear difference between sites. In the city the highest values occur after midnight. In the country they have a maximum at sunset and a minimum at sunrise. About 2 hr after sunrise the urban values drop notably, a condition which Hilberg labels "dry island." Obviously the evapotranspiration conditions are governing these processes more than anything else.

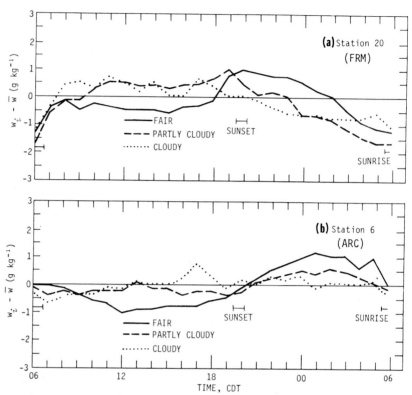

Fig. 8.1 Average diurnal variation of water-vapor mixing ratio (g kg^{-1}), based on observations during four summers at a rural (a) and an urban (b) station in the St. Louis area (from Hilberg, 1978).

This jibes with the assessment of Sisterson (1975) and Sisterson and Dirks (1978), who based their judgment on flights during the METROMEX project. Sisterson estimates that anthropogenic moisture production from all sources, as shown in Table 8.3, is only 1.1×10^9 g hr^{-1}. He compares this with a calculated potential evapotranspiration of about 6.7×10^{11} g hr^{-1} (in summer), that makes this component of the moisture balance overwhelming whenever water is available from precipitation or plant metabolism.

The flights above the city at about 750 m above the ground showed that dry regions correlated well with the intensity of the surface heat island. The "dry" regions correspond to residential areas and light industry land use. The dense settlement patterns of old dwelling

TABLE 8.3

**Moisture Production from Anthropogenic
Sources in St. Louis**[a]

Source	Amount (10^8 g hr^{-1})	Source	Amount (10^8 g hr^{-1})
Motor vehicles	2.3	Cement mills	1.5
Refineries	4.2	Steel mills	1.8
Power plants	1.0		

[a] After Sisterson (1975).

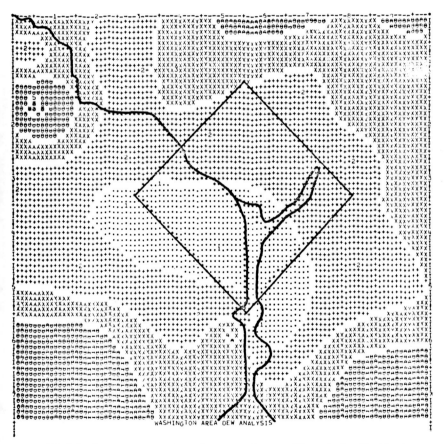

Fig. 8.2 Symap presentation of average dew conditions in the Washington, D. C., area for July 1974. Dark shading indicates heavy dew deposition, light stippling corresponds to light dew (from Myers, 1974).

units with lack of all vegetation were the driest areas. Quite notable in the overflights was the influence of the Mississippi River, where evaporation must be considerable.

All of this is corroborated by data collected by Myers (1974) on the dew deposits in summer in the Washington, D. C., metropolitan area. Average dew conditions for midsummer months are shown in Fig. 8.2. This pattern is probably representative of all urban areas. The reduced urban nocturnal cooling in the city results in less frequent lowering of temperature below the dew point or a much later occurrence of this event than in the countryside. There surfaces often cool below the dew point soon after sunset. In fact, Dzurisin (1978) reported for METROMEX that in St. Louis the dew points in the city are 0.5°C lower than in the surroundings. This, in addition to the reduced cooling rates, leads to less dew deposit than in the countryside. The reduction in dew is, in turn, a contributor to the heat-island formation. In the rural area the first radiative energy received during the day is used to evaporate the dew, but in the city it is immediately available to heat the surfaces and air, in the absence of dew.

8.2 URBAN CLOUDS

Cloud formation over urban areas is influenced both by the increased convection and by the enormous production of hygroscopic condensation nuclei. The former effect is more likely to produce clouds in summer, the latter is more apt to cause early condensation at inversion layers in the higher winter humidities. The formation of cumulus clouds in summer definitely starts earlier over the city than outside. This is clearly shown by the higher frequency of midday cloudiness at La Guardia Airport in New York City compared with the outlying Kennedy Airport (Table 8.4).

In the high-cloudiness class there is very little difference between the two airfields. But in the middle of the day there is a very notable shift in frequency at the urban field toward an increase, which lasts until the late afternoon. Of course, the anomalous cloud formation over industrial installations and cooling towers has been well docu-

TABLE 8.4

Midday Frequency of Various Levels of Cloudiness in July at La Guardia (LG) and Kennedy (K) Airports, New York, 1951–1960

	Cloud cover					
	0–0.3		0.4–0.7		0.8–1	
Hou	LG	K	LG	K	LG	K
1100	33	35	25	21	42	44
1300	24	31	35	27	41	42
150(	25	32	33	27	41	41
170(	31	31	26	24	43	45

mented (Koenig, 1979). Grosh (1977, 1978) noted that the first cumulus clouds in the St. Louis area are preferentially formed with three times the expected frequency over the central urban area and the refineries to the north of the city, which are sources of heat, water vapor, and nuclei.

In the METROMEX observations of Fitzgerald and Spyers-Duran (1973) at low-flight levels the cloud condensation nuclei (CCN) increased 54 percent from upwind to downwind of St. Louis for a supersaturation of 0.17 percent, and nearly doubled for a supersaturation of 1 percent. The more CCN that are present the smaller will be the cloud drops because of the increased competition for the available water vapor. These same authors were able to obtain a few drop-size distributions upwind and downwind from St. Louis. Their results showed that the average drop diameters were smaller downwind by about 2–3 μm and that a flat distribution of diameters changes to a steep, single-mode distribution. This is of considerable importance for initiation of precipitation, as we shall see in Section 8.3.

Semonin (1978a) reported on CCN at 300 m above ground over the METROMEX experimental area. The number concentration of CCN over the area, as presented in Fig. 8.3, shows that some of the high concentrations are related to known industrial sources. There is, in general, an order of magnitude increase in CCN from upwind of St. Louis to downwind. Measurements in clouds affected by CCN

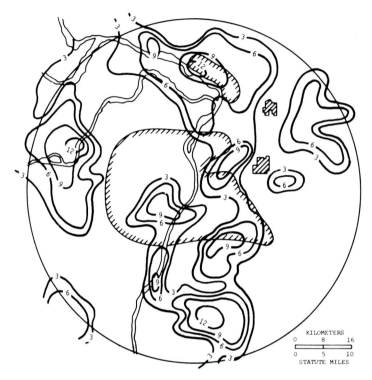

Fig. 8.3 Composite distribution of cloud condensation nuclei at 600 m above mean sea level observed over the St. Louis area during flights 1971–1975. Isolines give nuclei numbers of 10^4 cm^{-3} (from Semonin, 1978).

sources and unaffected, together with the liquid water content, is shown in Table 8.5 (Semonin, 1978b).

The increase in CCN over contaminating sources is highly significant but the slight increase in liquid water content is not. These numerical differences are not considered to be important for the initiation of precipitation, although chemical composition and hygroscopicity might be involved (Ochs and Semonin, 1977). Hygroscopic nuclei form quite rapidly in urban areas by catalytic oxidation of sulfur dioxide and hygroscopic sulfates. Oxides of nitrogen, too, can be transformed to nitric acid. Many very small particles or droplets result. Tomasi *et al.* (1975) measured in Milan, Italy, an industrial

TABLE 8.5

Cloud Condensation Nuclei (CCN) and Liquid Water Contents (LWC) in Low-Level Clouds over St. Louis in Summer

	CCN (cm^{-3})	LWC ($g\ m^{-3}$)
Clouds unaffected by CCN sources	1500	0.98
Clouds affected by CCN sources	3750	1.10

town, mass concentrations and size distributions indicative of available particulates for cloud formation (Table 8.6) during the heating season. This shows the abundance of small particles and the scarcity of large ones. If acting as CCN they can contribute to the colloidal stability of clouds.

A major role in the microphysics of clouds is played by ice nuclei, which can transform supercooled droplets to ice crystals. This change is often essential for the initiation of precipitation. The knowledge of ice nuclei (IN) in urban areas is extremely rudimentary and all statements about their action are essentially speculative. There are certain well-established sources but the published concentrations vary widely. One reason is that many of the observations were made from aircraft at quite different altitudes. Another reason is the lack of standardization of the various cloud chambers used and the use of different activation temperatures of these chambers for IN.

TABLE 8.6

Midwinter Particle Concentration in Milan, Italy[a]

Radius (μm)	Concentration	Radius (μm)	Concentration
0.15	4.9×10^2	0.75	1.4×10^1
0.20	6.5×10^2	1.0	2.27×10^0
0.25	5.7×10^2	1.5	5.5×10^{-1}
0.30	3.3×10^2	2.0	2.1×10^{-1}
0.40	2.0×10^2	2.5	9×10^{-2}
0.50	1.4×10^2	3	3×10^{-2}
0.60	6×10^2	4	1.5×10^{-2}

[a] Tomasi et al. (1945).

Among prolific sources of IN are steel mills. Telford (1960) measured in a plume over Newcastle, Australia, 300 IN liter^{-1}. Hobbs and Locatelli (1970) indicate that in the area of Seattle and Tacoma IN are in industrial sectors six times the background concentrations. In contrast, Braham and Spyers-Duran (1974) indicated that in 39 flights over St. Louis at 300 m there were 2 IN liter^{-1} upwind, 1.5 IN liter^{-1} over the city, and 1 IN liter^{-1} downwind (at activation temperatures down to $-16.6°C$). This decrease was attributed by these authors to deactivation over the city.

Another major source of urban ice nuclei was discovered by Schaefer (1968, 1969). Its origin is the lead in automobile exhaust interacting with traces of iodine to form lead iodine (Pb_2I), a very active ice nucleus. Small traces of iodine are always in air of maritime origin or result from combustion of organic materials. At ground level near exhausts of cars using leaded gasoline as many as 10^4 IN liter^{-1} were observed. On a busy highway Star (1975) observed between 1×10^3 to 3×10^4 IN liter^{-1}. More recently, Borys and Ducke (1979) have made some further analyses of Pb_2I in Providence, Rhode Island, where distinct contrasts between maritime and continental air masses exist. They found that of all trace metals studied, only lead had a significant positive correlation with the number of IN in the local atmosphere. A notable finding was also that the continental air-mass aerosol had about ten times less the surface area of the maritime aerosol. Lead particles from car exhaust are known to be particularly small and hence surface active. The concentration of IN at Providence, at $-17°C$ activation temperature, was generally below 100 liter^{-1}.

The state of knowledge of urban IN, their action on clouds, their life history and mass balance is very unsatisfactory. Until long-term observations about their nature and chemistry become available we shall remain in the dark about their effect on urban precipitation processes.

8.3 URBAN PRECIPITATION

Of all meteorological elements in the urban environment, precipitation still confronts us with the most puzzles. There are a number of reasons for this. Again the problem of topographic influence raises its ugly head. Precipitation amounts are extremely dependent on elevation and many urban areas have considerable height differences. The presence of water bodies that may stabilize air masses in summer and furnish moisture in winter has a great influence on the occurrence and amounts of precipitation. Yet in a very large literature there is much evidence for increases of precipitation in urban areas in comparison with rural environments near cities but not under their disturbing influences.

There are a number of factors that make an increase of precipitation by urban areas plausible. But these intermingle and their effects have not yet been disentangled. There are three main contributing causes for modification and augmentation of precipitation. The first and most obvious is, of course, the heat island. As already shown in Chapter 6, this leads to rising vertical motion over the cities and vertical motion is quintessential in the formation of precipitation. As can be demonstrated, this effect in the right combination with other weather conditions can initiate precipitation in urban areas.

The second cause for augmentation of precipitation is the obstacle effect. As already discussed, the aerodynamic roughness of urban structure impedes the progress of weather systems. If there are rain-producing processes taking place they may linger longer over the urban area and hence increase the amounts received at an urban location, compared with a rural spot where the front or occlusion moves faster. In many instances the obstacle and heat-island effects may combine, and it will be difficult to disentangle the contribution of each to the precipitation amounts.

The third factor affecting the precipitation are the pollution products. As we have seen these contribute to cloud formation and to changes in the drop-size spectra. As such they can either promote or inhibit precipitation. In case of ice nuclei, they might initiate the precipitation process in supercooled clouds. There are observations to show that all these processes do take place in urban areas, but the evidence is mounting that the pollution effect on urban precipitation is secondary to the aero- and thermodynamic impacts.

In all discussions about precipitation, urban or elsewhere, the high variability of this element and the sampling problem must be kept in mind. Also, the streakiness of rain or the passage of a single tropical storm can affect the statistics of amounts for a long time. This is the very reason why time-series analysis, which has been so helpful in determining urban trends in temperature, is not very fruitful for precipitation. Station moves, which have been frequent in urban areas, usually vitiate such attempts.

The method of comparing urban and rural stations is somewhat more fruitful, and in topographically uncomplicated settings has been used by many investigators. But even a glance at some of the results will show the difficulties to be had with this method. Table 8.7 shows a few urban–rural precipitation differences found in the literature.

None of the differences is larger than the standard deviation of annual precipitation. Thus statistically they cannot be very significant. The example of Chicago will illustrate this. In the interval 1951–1970, the mean annual precipitation there was 889 mm with a standard deviation of 197 mm and a range from 565 mm to 1170 m. In order to show or disprove urban influences on precipitation one has to resort to different methods of analysis. This has taken two paths. One is the scrutiny of network observations in and around urban areas. The other is to differentiate between rainfall events.

Looking at network data, albeit of relatively few rain gauges, Schmauss had already commented in 1927 that the maximum rainfall did not occur in the center of the city of Munich, but to the lee. In fact, he showed an increase from west to east, in accordance with the prevailing wind direction. He also showed that the surplus east of Munich was greater in summer than in other seasons, with a 15

TABLE 8.7

Urban–Rural Differences of Annual Precipitation Reported in the Literature

Locality	No. of years used	Precipitation (mm) Urban	Precipitation (mm) Rural	Difference (%)	Source
Moscow, U.S.S.R.	17	605	539	+11	Bogolopow, 1928
Urbana, Illinois	30	948	873	+9	Changnon, 1962
Munich, West Germany	30	906	843	+8	Kratzer, 1956
Chicago, Illinois	12	871	812	+7	Changnon, 1961
St. Louis, Missouri	22	876	833	+5	Changnon, 1969

percent increase over the western urban area. Changnon pioneered the network approach to urban precipitation monitoring, and in 1962 published a study of precipitation patterns over Urbana, Illinois. Isohyets for a 10-yr record are shown in Fig. 8.4. This has character- istics that are representative of similar studies. The amounts in- crease over the urban area, with a closed maximum to the lee of the city and then a decrease again. Even though Urbana is much smaller than Munich it is the same pattern observed by Schmauss. The METROMEX study for summer rainfall in St. Louis, in a summary analysis (Changnon, 1979), yielded similar conditions. A composite picture for all prerain wind directions showed for the total rainfall in 5 summers, with 302 rain events, a strong increase to the lee of the city (Fig. 8.5). The upwind–downwind difference is 23 percent, which is statistically significant. Even more startling was the in- crease at individual points in the 220-gauge network of MET- ROMEX. The area of Edwardsville showed a 49 percent enhance- ment of summer rainfall (Huff and Vogel, 1978). This was ascribed to the combined influence of St. Louis (distance, ~ 30 km) and the Altonwood River industrial complex (distance, ~ 8 km). Huff and Vogel attributed the increase to enhancement of the heavy rainfalls, yielding ≥ 25 mm. An excerpt from their tabulations is shown in Table 8.8.

Some very legitimate questions were raised about this analysis by Braham (1979). The knottiest problems are the so-called "upwind controls." Are those areas west and southwest of St. Louis really free of urban and topographic effects? And, in line with our earlier

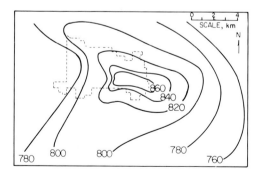

Fig. 8.4 Isohyets of annual precipitation (mm) at Urbana, Illinois (after Changnon, 1962). Dashed line marks urban limits.

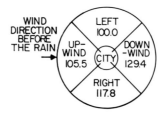

Fig. 8.5 Percent summer precipitation distribution in St. Louis area from METROMEX observations, shown in quadrants based on prerain wind direction (after Changnon, 1979).

comment on rainfall variability: Are five summers an adequate sample? Braham points out that frequency of large storms is more meaningful a measure than total rain amounts. There were, for example, 8 large storms observed upwind and 19 in Edwardsville. The ratio $19/8 = 2.38$ corresponds exactly to the claimed enhancement. The average storm frequency in the Edwardsville area was 111 compared with 120 in the upwind control area. Thus the rain increase in Edwardsville must have come from intensification of the storms that passed the city.

An independent verification of the Edwardsville rain enhancement was produced by Rao (1980), who subjected an early observation period (1910–1940) and a recent one (1941–1970) to trend analysis. The first interval was considered to be unaffected by the urbanization and industrialization processes and the second interval was

TABLE 8.8

Summer-Season Rainfall Distribution at Selected METROMEX Sites[a]

Site	Control, all events	Rainfall ratio to upwind area storms (≥25 mm)	Percent precipitation from storms (≥25 mm)	Diurnal extremes in 3-hr interval
Surburban, west	0.99	1.02	35	15–18
St. Louis:				
Urban	1.10	1.16	37	15–18
East, industrial	1.22	1.49	42	15–18
Wood River refineries	1.38	1.86	45	20–23
Edwardsville	1.49	2.37	55	14–17

[a] After Huff and Vogel (1978).

assumed to show the anthropogenic influence. Using an integrated moving-average model, the annual precipitation total showed an increase of 4.25 percent, significant at the 95 percent confidence level.

Long-term trends were also discovered downwind of the rapidly growing Tel-Aviv, Israel, area (Goldreich and Manes, 1979). Stations with three decades of record showed 5–17 percent rainfall increases, also statistically significant at the 95 percent confidence level. An intriguing result in this investigation was produced by a separate harmonic analysis of the annual variation of rainfall for two separate 30-yr intervals: 1901–1930, before urbanization and 1931–1960, during a period of rapid growth. The calculations showed a marked increase of November precipitation, i.e., at the beginning of the yearly rainy season in the area, and the first harmonic showed an advance of ~6 days of the maximum of rainfall. At 8 long-record stations, the November rainfall share was fairly consistently 12 percent of the annual total. It increased to 16 percent in the second period. The authors ascribe this to an urban effect on the relatively unstable air masses that prevail in the early part of the rainy season in the eastern Mediterranean.

A study of the precipitation in and near Turin, Italy, between 1952 and 1969 showed definite trends (Raffanelli and Papée, 1979). In that interval the city grew from 700,000 to 1.2 million inhabitants and the number of motor vehicles from 70,000 to 390,000. Summer-shower events showed an increase in frequency in the city but also an apparent decrease in precipitation volume per shower. In winter, also influenced by air from the industrial city of Milan, there seem to be decreases of both volumes and frequencies of precipitation in Turin. But in the adjacent rural zone there was a relative increase in precipitation events that is ascribed to nucleation by industrial aerosols. In another Italian city, Naples, Palumbo and Mazzarella (1980) found no precipitation trends in the interval 1886–1945. But in the past 30 yr, 1946–1975, with the growth and industrialization of the city an increase of 17 percent was noted.

An attempt to discover anthropogenic influences in the precipitation trends in four urban areas of New York State (Albany, Buffalo, New York City, and Syracuse) yielded essentially negative results, with the exception of New York City where a decrease in snowfall since 1940 was suggested (Jones and Jiusto, 1980).

Not all studies show increases in total rainfall amounts downwind. Sanderson and Gorski (1978) showed for the Detroit–Windsor area a decrease for the annual values in the interval 1970–1975. They point out that there are distinctly different seasonal patterns. In summer there was increased precipitation downwind but the greatest number of heavy precipitation days was in the western and southwestern sector of the urbanized area. It must also be noted that in a number of European cities the number of days with measurable precipitation is increased over the rural surroundings. In the colder season an increase of days with drizzle seems to be a distinct attribute of the industrialized cities of western Europe.

Another element that has obviously increased in urban areas are the socalled rain cells in summer. These are defined as cores of rain intensity. In the METROMEX study, Schickedanz *et al.* (1977) determined where such cells developed and where they traveled to in the urbanized area. Cell frequency in the surroundings were used as a control. Their most important findings are summarized in Table 8.9.

The implication one might read into these numbers is that the increases are related to pollutants and a great deal of effort has gone into investigating this hypothesis. In the St. Louis studies Huff and Schickedanz (1974) have presented estimates of the cloud condensation nuclei (CCN) production and the numbers shown in Table 8.10 are indeed startlingly large.

Gatz (1979) tried to correlate indices of aerosol-source strength from 48 sources with rainfall in 21 subareas of the St. Louis metropolitan area. He found only a few significant correlation coefficients,

TABLE 8.9

Summer Precipitation Produced by Rain Cells in the St. Louis area[a]

Sector	Area (km^2)	No. of cells	No. of cells (km^{-2})	Volume of rain (ha-m)[b]	Urban increase (%)
Control area	3576	4729	1.3	11.0	0
Bottom lands NNW of city	363	590	1.6	12.3	12
Hills, SW of city	375	436	0.9	19.1	74
Wood River industrial area	170	449	2.6	19.5	77
St. Louis:					
contiguous urban area	775	1746	2.2	24.8	125
industrial area	170	559	3.3	34.2	211

[a] From the METROMEX study.
[b] Hectare-meters.

TABLE 8.10

St. Louis Area CCN Production[a]

Sector	CCN active at		Supersaturation
	0.5 (%)	1 (%)	
Residential and downtown	~0.5	~0.7	$\times 10^4 cm^{-2} sec^{-1}$
Industrial area	1–2	2–3	$\times 10^4 cm^{-2} sec^{-1}$

[a] From Huff and Schickendanz (1974).

all explaining less than 36 percent of the variance. The study finds a relation to the refineries "suggestive" but ultimately concludes that the relatively weak agreement between the pollution indices and the rain areas in general might have included a few apparently significant correlations by chance.

The suggestion that the refinery effluents may, however, have a real influence on rain occurrence has recently been reinforced by a study in Los Angeles (Pueschel *et al.*, 1979). Airborne measurements of gases, aerosol-size distributions, and cloud systems yielded new data on the effect of refinery effluents. As species important for the microphysics of clouds, both sulfates and nitrates were found (and probably formed) in the refinery plumes. The nitrates are suspected of being the more active kind of nucleus. They are larger and more hygroscopic than the sulfates. The droplets forming on them have the tendency to be large (>1 μm). Their presence results in a wide dispersion of drop sizes in the cloud. This condition is well known to lead to further growth and rain, provided the cloud is thick enough. On the other hand, sulfates are very small particles (<0.1 μm) and more apt to stabilize clouds. In fact, the nonrefinery aerosol produced by the southern urban area of Los Angeles seems to have that effect. This shows again the great importance of the nature of the anthropogenic aerosol for the urban precipitation processes. Numbers alone will not explain the effects.

A fairly powerful argument for anthropogenic influences can be made when meteorological elements vary between the workweek and the weekend. This case was first made by Ashworth (1929), when he showed that in the industrial town of Rochdale rainfall had increased by 14 percent over three decades, but also that rainfall on Sundays was 13 percent less than on weekdays (154 mm versus 177 mm in the interval 1918–1927). A similar finding was published by Dettwiller (1970) for Paris, France for the interval 1960–1967. His results, reproduced in Fig. 8.6, show a gradual increase in average rainfall from Monday to Friday and then a sharp drop for Saturdays and Sundays. The weekend average was 1.47 mm, the workday average 1.93 mm; a decrease of 24 percent for the weekend. This study comprised 8 yr, but used only the warm season (May– October). A study by Frederick (1970) using 50 yr of data (1912–1962) at 22 urban stations in the eastern United Stated also noted a weekly cycle, but only in winter. The weekend values were about 8 percent less than the workday observations. But Frederick shows no weekly cycle for the warm season. We are again confronted by contradictions. A weekly cycle in winter may have some relation to the heating season. Mitchell (1961) noted in a comparison

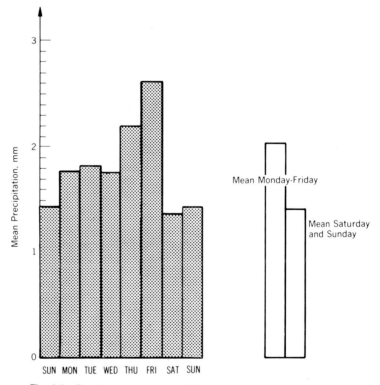

Fig. 8.6   Precipitation in Paris, France, averaged by day of week.

of airport versus urban temperatures in New Haven, Connecticut, that Sundays, in four winter seasons, showed the city only 0.3°C warmer. On weekdays the contrast of city–airport was 0.6°C for the daily mean temperatures. There might be more fuel used on weekdays than on Sundays in the urban workplaces and by vehicles.

The case of possibility of release of precipitation in industrialized areas cannot be closed out without some reference to the so-called LaPorte, Indiana, anomaly. This case, fortunately marked with a big question mark, was placed on record by Changnon (1968). In a 15-yr interval (1951–1965) rainfall was claimed to have increased 31 percent and a similar increase in thunderstorms was noted. The town itself is insignificant in size and activity, but it is 48 km east of the major steel-making and industrial district of South Chicago and Whiting, Indiana. That period was characterized by much production growth. The case immediately raised criticisms. A recording rain gauge in operation near the site of the regular station rain gauge since 1963 failed to show the increases claimed for the previous decade. Holzman (1971a) called the mechanism synoptically implausible. Hidore (1971), also crossing intellectual swords with Holzman (1971b), indicated that the anomaly was supported by Kankakee River runoff data. Tree-ring investigations by Ashby and Fritts (1972) were not able to resolve the difficulty but did not rule out that tree growth in the area was adversely affected by air pollution. Rao (1980) has taken up the case again by his statistical moving-average model. He verifies that during the interval 1929–1968, La Porte showed an increased rainfall trend compared to the historical record there. But a change of observers after 1967 seems to have eliminated the anomaly. In an epilog Changnon (1980) concluded that "the anomaly either ended or shifted into Lake Michigan," due possibly to changes in circulation patterns. The capriciousness of nature defies the observations from single rain gauges, even if properly taken, and thus LaPorte cannot be claimed as a support for the hypothesis that pollution increases local rainfall.

For the summer rainfall increases in urban areas an impressive case can be made for initiation by the heat island. There is a long list of papers reporting fairly isolated urban rainstorms and thunderstorms over urban areas. Schmauss (1927) had already shown a distinct surplus of heavy showers in the 20–40 mm class in the city of Munich compared with the suburbs. Berkes (1947) reported notably greater thunderstorm activity over Pest, Hungary, and Kratzer

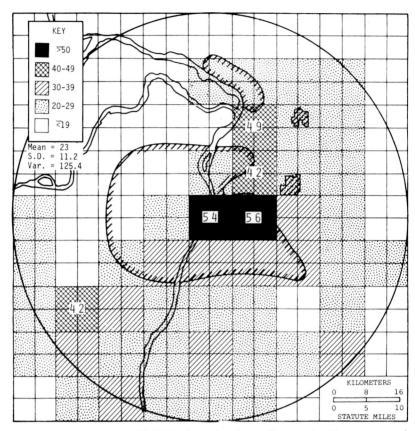

Fig. 8.7 Frequency of initial radar echo observations during the summers of 1972–1975 in the St. Louis area from the METROMEX project (from Huff, 1978).

(1956) indicated that Nürnberg, West Germany had 14 percent more thunderstorms in town compared with the airport.

Very suggestive is also Huff's (1978) frequency analysis of first radar echos in the St. Louis area in the 1972–1975 period. The distribution of 1550 initial echos shows (Fig. 8.7) a preponderance of central and industrial sectors of the urban areas. The existence of radar echos indicates the presence of large drops that start the rain process. Isolated summer showers over the Washington, D. C., urban area have been investigated by Harnack and Landsberg (1975). Figure 8.8 shows an example when a sector of the city received 25 mm from a single thunderstorm cell. There were no other

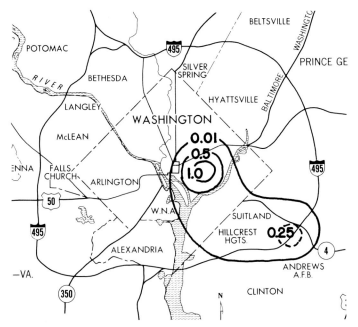

Fig. 8.8 Case of isolated rain shower in the Washington, D. C., area initiated by the urban heat island (isohyets in inches).

storms reported in the area. In a number of the cases the general synoptic situation did not warrant a rain forecast by the Weather Service and chances were generally rated below 20 percent. In these cases of isolated urban showers it could be shown that the ~2°C urban heat island was a trigger factor in the growth of cumulus clouds to cumulus congestus or cumulonimbus size with ensuing precipitation. In these cases geostrophic and higher level winds were usually weak, dew points were high, and very little impetus by thermal updrafts was needed to cause condensation and destabilization of the air mass. The added heat of condensation was thus sufficient to make the cloud grow to the precipitating state. Upper-wind translating moved these clouds often to the lee of the city. A box model was capable to calculate the lift of air parcels from the surface and their drift from the observed rawinsonde observations and, using an anticipated maximum urban temperature, predict whether or not or where urban showers were likely to form.

Occasionally the urban influence of lift by updraft will affect slowly moving, potentially unstable moist air masses or frontal zones. The result is city-centered, heavy shower activity. Almost every major city will show such cases when a major rain core stagnated over the urban area with only minor rainfall in the rural surroundings. Although these cases are mostly summer events, some occur in the transition season. An early account of such an anomalous occurrence was placed on record by Parry (1956) for Reading, England. This was an air-mass storm on June 22, 1951 drifting slowly through the region. Moisture was high to 3 km and instability was initially present from the surface to 2 km and from 4.5 to 8 km. A heat island of only 1°C was able according to Parry to trigger the storm, which dumped 34.5 mm rain on the city with only 2.5 mm recorded in the nearby rural area. In Washington, D. C., there are several well-documented cases of this type. Hull (1957) noted a storm of September 4, 1939 statistically as a once-in-a-hundred-years event, which inundated the city with 113 mm but yielded only 46 mm in the rural districts. A repetition on July 9, 1970 yielded 125 mm in the urban center but only 25 mm in neighboring rural Maryland (Landsberg, 1974).

The great synoptician Scherhag (1964) gave a graphic description of the cloudburst of June 27, 1964 in Berlin. With little upper motion a cumulonimbus formed over the city leading in 30 min to electrical discharges, rain, and hail. In 2 hr 81-mm rain fell, while to the southeast and northwest of the city almost no precipitation was observed. Another synoptic study of warm-season (May–September) urban releases of latent instabilities between 1961 and 1970 in Warsaw, Poland, was published by Lorenc (1978). Many of these intense storms involve stationary, occluded, and slow-moving warm fronts.

The most thorough synoptic investigations of urban storms have been carried out by Atkinson (1969, 1970, 1971, 1977). One of these storms originated over London, very similar to the case discussed by Scherhag. On August 21, 1959 at 1200 hr an intense heat island of both dry and wet bulb temperatures had developed over the city. This can be seen in a plot of the potential temperatures over the area (Fig. 8.9). The storm brought a maximum of 68 mm to the central urban area with no precipitation in the rural area at all. But Atkinson could also show that a megalopolis like London may have a profound influence on moving storms. His careful mesoscale analysis made this quite clear for the case of August 14, 1975. This was a day with fairly strong winds aloft: 10.8 m sec^{-1} at the 500-hPa level.

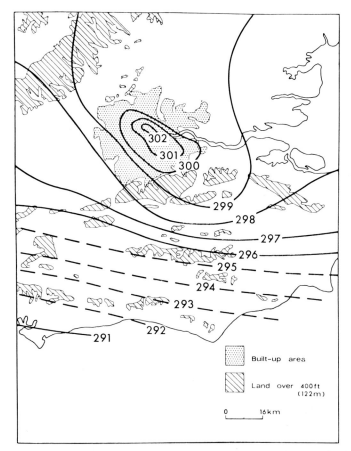

Fig. 8.9 Potential temperature (°K) distribution over London, England, prior to urban rainstorm, 1200 G.M.T., 21 August 1959 (from Atkinson, 1970).

There was a general mass convergence in the London region. This was reinforced by the urban heat island, which at 1200 hr was about 2.1°C warmer than the rural area. The vapor-pressure surplus of some areas in the city was an astonishing 6 hPa over the rural area. Clouds over the city started at 1300 hr. Shortly after 1600 hr some of the 200 recording rain gauges in the area caught the first precipitation. The heaviest downpours occurred between 1730 and 1800 hr. The isyhyetal pattern for a 24-hr period shows a very tight urban maximum, reaching 169 mm at Hampstead (Fig. 8.10) in the

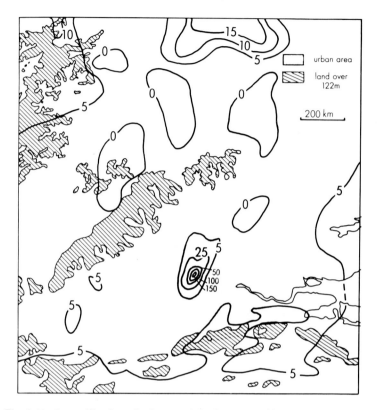

Fig. 8.10 Intensification of rain over the urban area of London, England (isohyets in millimeters) show 24-hr precipitation on August, 14–15, 1975 (from Atkinson, 1977).

northern part of the urbanized area. Many locations in the area had no rain at all and, more generally, only 5 mm fell in the area.

For planning purposes statistical estimates of heavy localized storm frequency are of great importance to civil engineers for drainage design. Wherever dense networks of recording rain gauges are available such design values can be calculated from a sufficiently long record. Vogel (1976) has done that, using the data from the Chicago network. In Fig. 8.9 is shown the pattern for the 2-hr storm rainfall to be expected statistically in a 5-yr interval. The 2-hr/5-yr rainfall is used by many United States municipalities for sewer design. Through the middle of the urban area 55–60 mm can be ex-

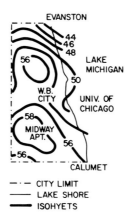

Fig. 8.11 Isohyets (mm) for the 2-hr 5-yr rainfall to be expected in the Chicago area (after Vogel, 1976).

pected that often, as seen in Fig. 8.11. Along the front of Lake Michigan, the values are somewhat lower, about 50 mm, probably due to the stabilizing influence of the water surface in summer. In the northern part, along the shore, in Evanston the values drop toward 40 mm ⎯⎯⎯⎯ ⎯⎯⎯⎯ difference in about 15 km is about 30 percent.

Although ⎯⎯⎯⎯⎯⎯⎯⎯ amounts and thunderstorm frequency se⎯⎯⎯⎯⎯⎯⎯⎯ downwind from the metropolitan centers, ⎯⎯⎯⎯⎯⎯⎯⎯ing observations. Heuseler (1964) noted ⎯⎯⎯⎯⎯⎯⎯⎯ y, the mean number of thunderstorm dav⎯⎯⎯⎯⎯⎯⎯⎯ ⎯urbs is 32 yr^{-1}, in the inner city only ?⎯⎯⎯⎯⎯⎯⎯⎯ he city 28 days yr^{-1}. Similarly, the sum⎯⎯⎯⎯⎯ ⎯⎯m in the southwestern approach zone ⎯⎯⎯⎯⎯⎯⎯ ⎯⎯ mm, in the inner city 160 mm, and in the north⎯⎯⎯ ⎯⎯⎯⎯⎯ ⎯an area 190 mm. There is no good explanation for the ⎯⎯⎯ent avoidance of the inner city by thunderstorms. But Fujita (1973) also noted a curious avoidance of central Chicago and central ⎯⎯⎯⎯ by tornadoes. This noted tornado expert reported that "a

shaped tornado area around Tokyo. In his opinion the heat island of the city is so uniform that significant vorticity is mainly generated at the interface with the cooler country air. In addition the increased friction at the urban surface is assumed to decrease tornado inten-

TABLE 8.11

Maximum Urban–Rural Differences in Summer Rainfall and Severe Weather Events, Expressed as Percent of Rural Values, and Based on Available Historical Data[a]

City	Rainfall	Thunderstorms	Hailstorms
St. Louis	+15	+25	+276
Chicago	+17	+42	+246
Cleveland	+27	+38	+90
Indianapolis	0	0	0
Washington, D. C.	+9	+36	+67
Houston	+9	+10	+430
New Orleans	+10	+27	+350
Tulsa	0	0	0
Detroit	+25	No data	No data

[a] From Changnon (1976).

sity. The tornado-suppressing elements of large metropolitan areas need still further exploration.

Changnon (1976) summarized the summer precipitation and storm increases in the lee area of major United States cities. His somewhat contradictory results are shown in Table 8.11. In view of the extraordinary variability of hail incidence the indicated increases must be met with considerable scepticism. Changnon related the thunderstorm increases to population as shown in Fig. 8.12.

It remains to take a look at winter precipitation in climates where snow occurs. Two phenomena require attention. One is a direct effect of the heat island. While it snows in the countryside, the city often has only rain. A contributing factor to this is the fact that snow will occur with high frequency when the surface temperature is close to or slightly below freezing. Kassner (1917) noted that in Berlin the relative frequency of precipitation falling in form of snow decreased in the city. When it snowed in the countryside it snowed in only 72 percent of the cases also in the city, in 14 percent of the cases the urban rain was mixed with snow, in 7 percent only rain was noted in the city, and in the remaining 7 percent the city had no precipitation. Paris, with very few days of snow per year has about 30 percent fewer cases than the nearby rural area (Maurain, 1947). In New York City, the long-term rural snowfall is 91 cm per season, but in the city in Central Park only 76 cm fall, a 16.5 percent reduction. Grillo and Spar (1971) on the basis of observations from 33 stations

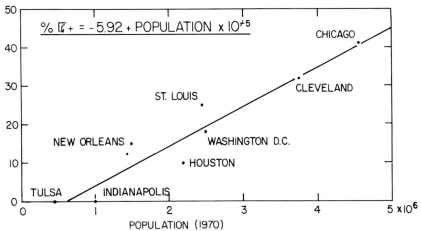

Fig. 8.12 Percent increase in thunderstorm incidence as related to urban population size (after Changnon, 1976).

concluded that snow probability when synoptic conditions favor precipitation in winter in the New York area is 40–45 percent in the rural areas but only 27–35 percent in the city. In Chicago the suburban Midway Airport has an average of 84 cm annual snowfall, but the downtown Loop area receives only about 65 cm, a 23 percent reduction.

The urban effect was beautifully demonstrated by Lindquist (1968) for a case in Lund, Sweden. Measurements of snow depth during a storm showed a systematic decrease of snow amounts from 8 cm in rural areas to 3 cm in the urban center, as shown in Fig. 8.13. The lines of equal snow depth are a reverse image of heat-island isotherms. In this case, the urban area was nearly snow free on the day following the storm, whereas the countryside had a solid snow cover.

The second winter phenomenon deals with urban-induced precipitation. Although there is no statistically significant evidence that total winter precipitation is appreciably above the rural values, there are definite indications that the number of days with drizzle or traces of snow is larger in the cloudy winter climate of the western European industrial cities than in the surroundings. Kratzer (1937) reported on winter trips from the outskirts to the center of Munich visual observations of snowfall from fog or stratus over the city but none in the surroundings. We have already seen that some industrial

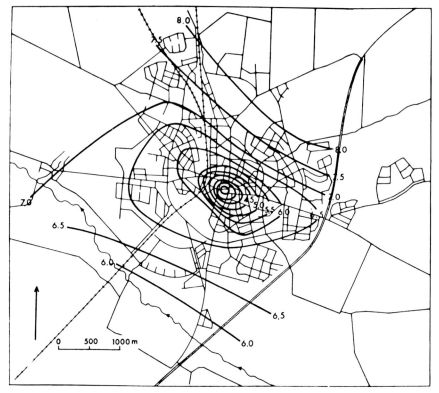

Fig. 8.13 Effect of urban area of Lund, Sweden, on snowfall. Lines of equal snow depth in millimeters (from Lindquist, 1968).

processes and car exhaust are sources of ice nuclei, active at various temperatures. Supercooled low-stratus clouds are quite common in winter in various higher latitude industrial areas. There is compelling evidence that these are occasionally nucleated and produce light snow. A convincing case study was presented by von Kienle (1952). He observed on January 28–29, 1949, over the heavily industrialized cities of Mannheim and Ludwigshafen a snowfall from a 500-m-thick layer of fog and low stratus, which was restricted to the urban area. Nearby Heidelberg had neither clouds nor snow. In Mannheim, with calm weather at −4°C, a light snow dusting of about 6 mm was noted.

A similarly anomalous snowfall was reported by Culkowski (1962) at the gaseous diffusion plant, 15 km southwest of Oak Ridge, Tennessee, in the cooling tower plume of the plant. About 5–8 km from the cooling tower the effluent had formed a low stratus. No other clouds were in the area. The clouds were definitely supercooled, with temperatures between -12 and $-16°C$ and slight winds of 3 m sec^{-1}. Ice nuclei from a ferromanganese plant, 29 km upwind were suspected as the active agents. A snow depth of 2.5 mm covered an area of 6.5 km^2.

On January 11, 1971 a similar case of light snowfall was observed by Agee (1971). Ice nuclei from the stacks of a coal-fired power plant were suspected to have induced a 6-mm snowfall in a supercooled fog. The temperature was $-12.8°C$ and aluminum oxide found in the stack effluent, active at $-6.5°C$ was suspected to have nucleated the droplets.

A very similar case was observed near Boulder, Colorado (Parungo et al., 1978). A coal-fired power plant there produces ice nuclei, some of which are already effective at $-5°C$. On a winter day a snowfall was clearly linked to the plume. Outside the plume there were 5 IN liter^{-1}, in the plume 200 IN liter^{-1}. These seeded a cloud and caused under the 0.5-km-wide plume a snow cover of 3 cm up to 5 km downwind and 1–2-cm snow to 10–20 km downwind.

The chain of events in these isolated cases, although not ironclad, seems to be fairly straightforward and convincing. More complex, undoubtedly, are enhanced seasonal snowfalls in and to the lee of urban areas. One has to assume persistent synoptic conditions causing supercooled low-level clouds and continuously functioning sources of ice nuclei. A suspicious case of this nature was noted during the 1959–1960 cold season in and near Toronto, Ontario. The snow pattern was quite at variance with the usual distribution, which shows an increase from Lake Ontario northward, ranging on an average from 100 to 140 cm. In the anomalous year Potter (1961) noted that the heaviest snowfall of 211 cm was in the center of the city. A plumelike formation of lines of equal snowfall extended from the west end of the city eastward to the lee, with decreasing total values toward the lake in the south and the countryside in the north (Fig. 8.14). Such patterns, observed after the fact, may be suggestive of an urban influence, possibly a combination of nucleation and heat-island effects, but they are far from a scientifically satisfactory proof.

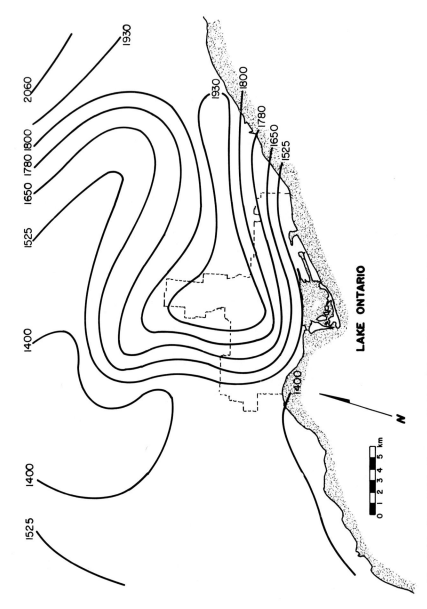

Fig. 8.14 Snowfall in Toronto, Ontario (Canada), region during winter 1959–1960 (mm). Dashed line shows limits of urban area (after Potter, 1961).

References

Agee, E. M. (1971). An artifically induced local snowstorm. *Bull. Am. Meteorol. Soc.* **52,** 557–560.

Ashby, W. C., and Fritts, H. C. (1972). Tree growth, air pollution, and climate near LaPorte. *Bull. Am. Meteorol. Soc.* **53,** 246–251.

Ashworth, J. R. (1929). The influence of smoke and hot gases from factory chimneys on rainfall. *Q. J. Roy. Meteorol. Soc.* **55,** 341–350.

Atkinson, B. W. (1969). A further examination of the urban maximum of thunder rainfall in London, 1951–1960. *Inst. Brit. Geogr. Trans. Papers* **48,** 97–117.

Atkinson, B. W. (1970). The reality of the urban effect on precipitation, a case study approach. *In* "Urban Climates," *WMO Tech. Note,* No. 108, pp. 342–360.

Atkinson, B. W. (1971). The effect of an urban area on the precipitation from a moving thunderstorm. *J. Appl. Meteorol.* **10,** 47–55.

Atkinson, B. W. (1977). Urban effects on precipitation: An investigation of London's influence on the severe storm in August 1975. *Occasional Papers,* No. 8, 31 pp. Dept. of Geography, Queen Mary College, Univ. of London.

Berkes, Z. (1947). A Csapadek eloszlasa Budapest teruleten. *Időjaras* **51,** 105–111.

Bogolopow, M. A. (1928); Über das Klima von Moskau. *Meteorol. Z.* **45,** 152–154.

Borys, R. D., and Ducke, R. A. (1979). Relationship among lead, iodine, trace materials and ice nuclei in a coastal urban atmosphere. *J. Appl. Meteorol.* **18,** 1490–1494.

Braham, R. R. (1979). Comments on urban, topographic and diurnal effects on rainfall in the St. Louis region. *J. Appl. Meteorol.* **18,** 371–375.

Braham, R. R., and Spyers-Duran, P. (1974). Ice nucleus measurements in an urban atmosphere. *J. Appl. Meteorol.* **13,** 940–945.

Changnon, S. A., Jr. (1961). Precipitation contrast between the Chicago urban area and an offshore station in sourthern Lake Michigan. *Bull. Am. Meteorol. Soc.* **42,** 1–10.

Changnon, S. A., Jr. (1962). A climatological evaluation of precipitation patterns over an urban area. *In* "Air Over Cities," *SEC Tech. Rept.* A62-5, pp. 37–66. U. S. Public Health Serv., Cincinnati, Ohio.

Changnon, S. A., Jr. (1968). The LaPorte weather anomaly: Fact or fiction? *Bull. Am. Meteorol. Soc.* **49,** 4–11.

Changnon, S. A., Jr. (1969). Recent studies of urban effects on precipitation in the United States. *Bull. Am. Meteorol. Soc.* **50,** 411–421.

Changnon, S. A., Jr. (1976). Inadvertent Weather Modification. *Water Res. Bull.* **12,** 695–718.

Changnon, S. A., Jr. (1979). Rainfall changes in summer caused by St. Louis. *Science* **205,** 402–404.

Changnon, S. A., Jr. (1980). More on the LaPorte Anomaly: A Review. *Bull. Am. Meteorol. Soc.* **61,** 702–717.

Culkowski, W. M. (1962). An anomalous snow at Oak Ridge, Tennessee. *Mo. Wea. Rev.* **90,** 194–196.

Dettwiller, J. (1970). Incidence possible de l'activité industrielle sur les precipitations à Paris. *In* "Urban Climates," *WMO Tech. Note,* No. 108, pp. 361–362.

G. Dzurisin (1978). Average dew points. *In* "Summary of METROMEX," Vol. 2. *Ill. State Water Survey Bull.* 63, pp. 43–62.

Fitzgerald, J. W., and Spyers-Duran, P. A. (1973). Changes in cloud nucleus concentration and cloud droplet distribution associated with pollution from St. Louis. *J. Appl. Meteorol.* **12**, 511–516.

Frederick, R. H. (1970). Preliminary results of a study of precipitation by day-of-week over the eastern United States. *Bull. Am. Meteorol. Soc.* **51**, 100.

Fujita, T. T. (1973). Tornadoes around the world. *Weatherwise* **27**, 56–62, 79–83.

Gatz, D. F. (1979). An investigation of pollutant source strength-rainfall relationships at St. Louis. *J. Appl. Meteorol.* **18**, 1245–1251.

Goldreich, Y., and Manes, A. (1979). Urban effect on precipitation on patterns in the greater Tel-Aviv area. *Arch. Met. Geophys. Biokl. Ser. B.* **27**, 213–224.

Grillo, J. N., and Spar, J. (1971). Rain–snow climatology of the New York metropolitan area. *J. Appl. Meteorol.* **10**, 56–61.

Grosh, R. (1977). Satellite observed urban cloud distributions. *Conf. Inadvertent Planned Weather Modification, Champaign-Urbana, 6th,* pp. 45–48. Preprints, Am. Meteorol. Soc., Boston.

Grosh, R. C. (1978). Studies of anomalous cumulus clouds. *In* "Summary of METROMEX," Vol. 2. *Ill. State Water Survey Bull.* 63, pp. 212–228.

Harnack, R. P., and Landsberg, H. E. (1975). Selected cases of convective precipitation caused by the metropolitan area of Washington, D. C., *J. Appl. Meteorol.* **14**, 1050–1060.

Heuseler, H. (1964). Zur unterschiedlichen Verteilung der Gewittertage im Berliner Raum. *Wetter and Leben* **16**, 197–203.

Hidore, J. F. (1971). The effects of accidental weather modification on the flow of the Kankakee River. *Bull. Am. Meteorol. Soc.* **52**, 99–103.

Hilberg, S. D. (1978). Diurnal temperature and moisture cycles. *In* "Summary of METROMEX," Vol. 2. *Ill. State Water Survey Bull.* 63, pp. 25–42.

Hobbs, P. V., and Locatelli, J. D. (1970). Ice nucleus measurements at three sites in Western Washington. *J. Atmos. Sci.* **27**, 90–100.

Holzman, B. G. (1971a). LaPorte precipitation fallacy. *Science* **171**, 847.

Holzman, B. G. (1971b). More on the LaPorte fallacy (with reply by J. F. Hidore). *Bull. Am. Meteorol. Soc.* **52**, 572–574.

Huff, F. A. (1978). Radar analysis of urban effects of rainfall. *In* "Summary of METROMEX," Vol. 2. *Ill. State Water Survey Bull.* 63, pp. 265–273.

Huff, F. A., and Schickedanz, P. T. (1974). METROMEX: Rainfall analysis. *Bull. Am. Meteorol. Soc.* **55**, 90–92.

Huff, F. A., and Vogel, J. L. (1978). Urban topographic and diurnal effects on rainfall in the St. Louis region. *J. Appl. Meteorol.* **17**, 565–577.

Hull, B. B. (1957). Once-in-hundred-year rainstorm, Washington, D. C., 4 September 1939. *Weatherwise* **10**, 128–131, 139.

Jones, P. A., and Jiusto, J. E. (1980). Some local climate trends in four cities in New York State. *J. Appl. Meteorol.* **19**, 135–141.

Kassner, C. (1917). Der Einfluss Berlins als Groszstadt auf die Schneeverhältnisse, *Meteorol. Z.* **34**, 136–137.

Kienle, J. von (1952). Ein stadtgebundener Schneefall in Mannheim. *Meteorol. Rundsch.* **5**, 132–133.

Koenig, L. R. (1979). Anomalous cloudiness and precipitation caused by industrial heat rejection. *Rand Corp. Rept.* R-2465-DOE, 100 pp, Stanta Monica, California.

Kratzer, A. (1937). "Das Stadtklima," Die Wissenschaft, Vol. 90 (1st ed.), 145 pp. Friedr. Vieweg & Sohn, Braunschweig.

Kratzer, A. (1956). "Das Stadtklima," Die Wissenschaft, Vol. 90 (2nd ed.), 184 pp. Vieweg & Sohn, Braunschweig.

Landsberg, H. (1974). Inadvertent atmospheric modification through urbanization. In "Weather and Climate Modification," (W. M. Hess, ed.), pp. 754–755. Wiley, New York.

Landsberg, H. E., and Maisel, T. N. (1972). Micrometeorological observations in an area of urban growth. Boundary Layer Meteorol. 1, 61–63.

Lindquist, S. (1968). Studies on the local climate in Lund and its environs. Lund Studies, Geogr. Ser. A 42, 79–93.

Lorenc, H. (1978). Opady Ulewne i nawalne na Obzarze Wielkiej Warszawy (Heavy showers and rainstorms in the area of greater Warsaw—translated title). Przeglad Geofiz. 23 (4), 271–294.

Maurain, C. H. (1947). "Le Climat parisien," 163 pp. Presses Univ., Paris.

Mitchell, J. M., Jr. (1961). The temperature of cities. Weatherwise 14, 224–229, 258.

Myers, T. M. (1974). Dew as a visual indicator of the urban heat island of Washington, D. C., MS Thesis, 54 pp., Univ. of Maryland, College Park.

Ochs, H. T., III, and Semonin, R. G. (1977). The sensitivity of cloud microphysics to an urban environment, Conf. Inadvertent Planned Weather Modification, Champaign-Urbana, 10–13 Oct. 1977, 6th pp. 41–44. Am. Meteorol. Soc., Boston.

Palumbo, A., and Mazzarella, A. (1980). Rainfall statistical properties in Naples. Mo. Wea. Rev. 108, 1041–1045.

Parry, M. (1956). An "urban rainstorm" in the Reading area. Weather 11, 41–48.

Parungo, F. P., Allee, P. A., and Weickmann, H. K. (1978). Snowfall induced by a power plant plume. Geophys. Res. Letters 5, 515–517.

Potter, J. G. (1961). Areal snowfall in metropolitan Toronto. Rept. CiR-3431 TEC-342, 9 pp. Meteorol. Branch, Dept. of Transport., Toronto.

Pueschel, R. F., Parungo, F. P., Barrett, E. W., Wellman, D. L., and Proulx, H. (1979). Meteorological effects of oil refinery operations in Los Angeles. Natl. Oceanic Atmos. Adm. Tech. Mem., ERL APCL-22, CO, 62 pp.

Raffanelli, C. E., and Papée, H. M. (1979). Long-term relative trends of precipitation over an industrial city and some adjacent rural areas of Piedmont. Riv. Ital. Geofis. Sci. Affini 5, 163–166.

Rao, A. R. (1980). Stochastic analysis of annual rainfall affected by urbanization. J. Appl. Meteorol. 19, 41–52.

Sanderson, M., and Gorski, R. (1978). The effect of metropolitan Detroit–Windsor on precipitation. J. Appl. Meteorol. 17, 423–427.

Schaefer, V. J. (1968). Ice nuclei from auto exhaust and organic vapor. J. Appl. Meteorol. 7, 148–149.

Schaefer, V. J. (1969). The inadvertent modification of the atmosphere by air pollution. Bull. Am. Meteorol. Soc. 50, 199–208.

Scherhag, R. (1964). Der Berliner Siebenschläfer Wolkenbruch. Berliner Wetterkarte, Beil. 74/64, SO 34/64.

Schickedanz, P. T., Busch, M. B., and Green, G. D. (1977). METROMEX raincell studies 1971–1975. Conf. Inadvertent Planned Weather Modification, Champaign-Urbana, 6th, pp. 57–60. Preprints Am. Meteorol. Soc., Boston.

Schmauss, A. (1927). Groszstädte und Niederschlag. *Meteorol. Z.* **44**, 339–341.

Semonin, R. G. (1978a). Aerosol patterns. *In* "Summary of METROMEX," Vol. 2. *Ill. State Water Survey Bull.* 63, pp. 103–104.

Semonin, R. G. (1978b). Cloud characteristics. *In* "Summary of METROMEX," Vol. 2. *Ill. State Water Survey Bull.* 63, pp. 236–239.

Sisterson, D. L. (1975). Studies on the urban moisture budget. Rept. No. AS114, 52 pp. Dept. Atmos. Sci., Univ. of Wyoming, Laramie.

Sisterson, D. L., and Dirks, B. A. (1978). Structure of the daytime urban moisture field. *Atmos. Environ.* **12**, 1943–1949.

Star, A. M. (1975). The design of a cloud chamber and its application in determining ice nuclei concentrations. M.S. Thesis, 47 pp. Univ. of Maryland, College Park.

Telford, J. W. (1960). Freezing nuclei from industrial processes. *J. Meteorol.* **17**, 676–679.

Tomasi, C., Guzzi, R., and Vittori, O. (1975). The "SO_2-NH_4—solution droplets" system in an urban atmosphere. *J. Atmos. Sci.* **32**, 1580–1586.

Vogel, J. L. (1976). Heavy rainfall in a major metropolitan area. *Conf. Hydrometeorol., April 20–22, 1976, Ft. Worth, Texas,* pp. 86–91. Preprints Am. Meteorological Society, Boston.

Zanella, G. (1976). Il clima urbano di Parma. *Riv. Meteorol. Aeronaut.* **36**, 125–146.

9

Urban Hydrology

Many human settlements are located near water bodies. Aside from their esthetic attractions, streams, lakes, and rivers became the location of villages, towns, and cities because they were a source of water supply. The rivers offered an early means of transportation and their valleys made construction of highways and railroads easier. Unfortunately, the running water also seemed, in many cases, to offer a cheap way of sewage disposal. Artificial lakes and reservoirs for beautification and water supply are also part of many urban scenes.

We have already seen the notable changes in the urban hydrological cycle, especially the apparent increases in rainfall. Added to this are radical changes in surface characteristics in urban areas, leading to rapid runoff of precipitation and flooding. The problem of water management in cities has become a major problem of modern society. Poor practices in land utilization have led to large economic losses. Designs of drainage systems pose major engineering tasks and deterioration of water quality has presented health concerns. Urban hydrology has become therefore, a rapidly growing interdisciplinary field. A recent review of work in the last half-decade lists no fewer than 217 references (McPherson, 1979), and the first textbook exclusively devoted to this topic has appeared (Lazaro, 1979). Our concern here will be primarily with the meteorological aspects.

Urban floods constitute one of the principal weather hazards. Loss of life and damages are often very high. Although a whole watershed will contribute to the flooding, urbanized areas vastly aggravate the trouble. The reason is the same as one of the principal causes of the heat island. The radical change of surface from permeable soils and vegetation to impermeable pavements, parking lots, and roofs shortens the time rain or meltwater reaches the water courses. Storm drain systems deliberately designed to carry the water away from residential and business districts reduce the lag times even further. The benefit derived by rapidly drying-off traffic arteries sometimes leads to the high cost of flooding.

Natural surfaces permit rainfall to penetrate the soil, where some of the water is stored and gradually infiltrates into the groundwater table. Even in heavy rainfalls surface runoff is considerably retarded. In vegetated areas much of the water is temporarily intercepted by leaves, needles, and trunks. Some of it is evaporated by the plants. The forest litter is a particularly effective storage medium. Thus it is clear that the runoff in heavy storms is governed by the ratio of permeable to impermeable ground cover. This ratio is not too readily arrived at. It is completely empirically determined either from aerial photographs or from satellite pictures. Other methods attempt to derive a numerical value for impervious area from land use and population density. The general form of such a relation is

$$I = aD^b$$

where

I percentage of impervious land
D population density
a, b constants dependent on land use

The constants in such a formula are determined by calculating the contribution of each land use in a multiple regression relation to imperviousness. The classification used generally includes the percentages of forest, grassland, agricultural land, residential housing, commercial buildings, parking lots, streets, and highways.

The percentage of impervious area is then related to the water discharge in the streams or the river of the drainage basin in which the community is located. The discharge is usually depicted by a flood hydrograph that indicates the volume of water passing the gauging point per unit time. Figure 9.1 shows in a highly schematized way the change undergone by the hydrograph for a hypothetical intense rainfall for a basin changing from natural surface to a sewered basin with a highly impervious surface. It clearly shows the relatively slow rise of stream discharge in a rural area with natural surface, where runoff is slow and part of the rain is stored in soil and groundwater, and some evaporates. There the flood discharge subsides slowly and the peak flow is moderate. In contrast, in an area with impervious ground cover, the rise in discharge rate is precipitous to a very high peak and a rapid decline too. It is the sudden surge that leads to the extensive damage in low-lying urban areas.

How closely this hypothetical hydrograph resembles a real situa-

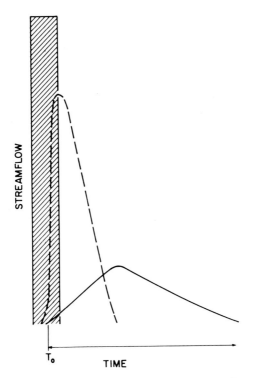

Fig. 9.1 Hypothetical streamflow curves in urban (dashed) and rural (solid) areas as result of intense rainfall (hatched bar). (Adapted from Anderson, 1970).

tion can be illustrated by an observed case in a newly urbanized area in Maryland. One residential sector of the new town of Columbia with about 30 percent impervious area and good storm sewers was drained by a little stream. Every shower led to almost immediate rise of the water level in the stream. A typical example is shown in Fig. 9.2. The rain shield of a coastal low brought showery precipitation with 19.3 mm in about 3 hr. This led to an almost instantaneous rise of 27.4 cm in the stream (Landsberg, 1979).

Much effort has been devoted to obtain measures of the peak discharge as function of rain amount and of the time lag between the time of rain occurrence and the time of this peak discharge. Again, the relations are strictly empirical and statistical. Each urban drainage area has its own characteristics. In generalized form

$$Q_p = K i_p A \qquad (9.1)$$

where

Q_p peak runoff rate (m³ sec⁻¹)
K dimensionless runoff coefficient
i_p rainfall intensity (mm hr⁻¹)
A drainage area (hectares)

Values for the coefficient K are not too precisely known, but Table 9.1 gives ranges for K that are conventionally used.

By estimating the percentage of various land-use categories in an

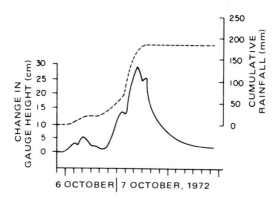

Fig. 9.2 Stream levels due to runoff in urban area (solid line, left scale) following rainfall (recording rain gauge level, dashed line, right scale). Time on abscissa in hours. (From Columbia experiment, Landsberg, 1979.)

TABLE 9.1

Runoff Coefficients for Urban Areas

By surface type (for flat areas)	K values
Lawns:	
Sandy soil	0.05–0.10
Loamy soil	0.10–0.20
Park areas	0.10–0.25
Asphalt or concrete surfaces	0.70–0.95
Roof areas	0.75–0.95

By land use	K values
Unimproved	0.10–0.30
Suburban residential	0.25–0.40
Single family units (urban)	0.30–0.50
Apartments	0.50–0.70
Business section (central city)	0.70–0.95
Industry:	
light	0.50–0.80
heavy	0.60–0.90

urban area and, giving each corresponding runoff coefficient a proportionate weight, a combined runoff coefficient can be obtained:

$$K_n = aK_1 + bK_2 + \cdots + nK_n \qquad (9.2)$$

where the coefficients $a, b, \ldots, n$ represent the percent share of each land-use category of the whole area.

For sloping surfaces the coefficients have to be increased. The magnitude of the increase depends on the degree of imperviousness and may range from 0.10 to 0.30.

The time lag for the maximum discharge has been also variously expressed, with a common form being

$$\tau = f(\log d/\sqrt{s})$$

where d is the distance of the water course from runoff area and s a slope index. Actual values for the function f are again empirically developed for each case (Anderson, 1970).

For many design purposes it is desirable to make some estimates of discharges (or flood heights) and time lags for varying rainfall amounts in specific time intervals. Such estimates have to be made on the basis of available rainfall records. Because the worst flood situations can arise from very short, sharp thunderstorms, hence only where an adequately long period of recording rain-gauge data is available can these estimates be made. In most instances only 24-hourly rainfall values are at hand for frequency analysis. Fortunately these cover such extreme amounts as are experienced in tropical storms.

The practice is to derive from the rainfall, again empirically, the maximum rainfall amounts expected in a given recurrence interval of 2, 5, 10, 20, 50, or 100 yr. If the series is shorter than the longer intervals, these values are estimated by use of a suitable statistical analysis of extreme values. The recurrence intervals are generally labeled as "return periods" or as the 20-yr or 50-yr storms. It must be clearly understood here that these designations do not imply that such events occur even approximately at equal intervals in time. There is nothing to prevent two so-called 50-yr storms from occurring in adjacent years.

A large number of estimates have been made for various areas of the world for the maximum floods to be expected (Gray, 1973). These take in smaller basins the form

$$Q_{max} = mA^y$$

where A is the area and m and y are empirical constants.

In graphical form a curve representing such a relation for a city-size area or its subdivisions is shown in Fig. 9.3. The values in use can actually vary by orders of magnitude. In an urban area that slopes steeply toward a river it is obviously greater than in flat terrain for equal amounts of rain. In Fig. 9.4, a hypothetical example of ratios of runoff in a rural area to an urbanized area of varying imperviousness is depicted for a 2-yr rainfall on 8 hectares (Terstriep *et al.*, 1976). Obviously similar diagrams can be developed for other return periods. On the figure a parameter called concentration time is also shown. This gives a time dimension comparable to the time lag for peak runoff. The dimension of the runoff area is quite small, only a few city blocks in size, with an example of runoff in an ordinary rainfall shown in Fig. 9.5. One can piece larger areas together from a mosaic of small urban portions with uniform land use. This has an advantage because rarely is rainfall, especially in thunderstorms, uniform over a whole city. We will refer to this again later.

Another way to depict the effect of varying degrees of varying imperviousness for larger basins is shown in Fig. 9.6. This shows the

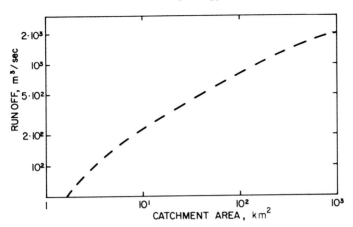

Fig. 9.3 Runoff in relation to catchment area from a simplified empirical model. (Adapted from Gray, 1973.)

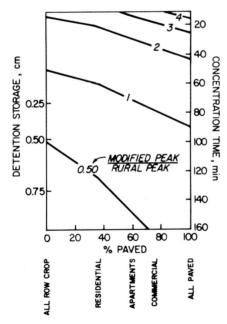

Fig. 9.4 Ratio of runoff peak urban/rural in relation to land use (solid curves). (Adapted from Terstriep *et al.*, 1976.)

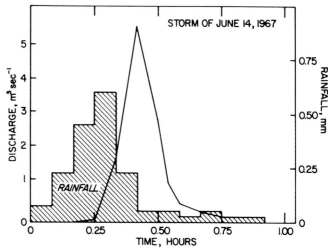

Fig. 9.5 Storm discharge (solid line) in small catchment area as a result of short-duration rainfall (hatched column). (Adapted from Terstriep *et al.*, 1976.)

ratio of flood size for various recurrence intervals to the average flood that corresponds to a 2.3-yr flood (Anderson, 1970).

Analyses of flood potential in larger cities based on single rain gauges, or on Weather Service stations at airports at some distance from the city are simply inadequate. Good illustrations for the accuracy of this statement can be obtained from the detailed rain-gauge data collected by project METROMEX in St. Louis. Huff and Vogel (1977) have published maximum values of rainfall for 5-min and 1-hr intervals from a network of 223 recording rain gauges in a circular area of about 5180 km². Their tabulations cover primarily intense storm cells either occurring as thunderstorms or as embedded convective cells in larger storm systems. The differences of rain amounts received are startling. On August 28, 1974 one gauge in the network recorded 1.02 in. (25.9 mm) in 5 min, but it was another gauge that received the largest hourly amount during that storm of 2.52 in. 64.0 mm). Rainfall in that case came from a slow-moving frontal system and the largest amounts were received in the northerly sectors of the urban area. In another incident, during the passage of a squall line on July 9, 1973 one rain gauge recorded at 1600 hr 3.02 in. (76.7 mm) in 1 hr. But the next-largest amount at a nearby gauge was only 1.43 in. (36.3 mm) for the same hour. In that case the highest rainfall values occurred in the southern sectors of the metro-

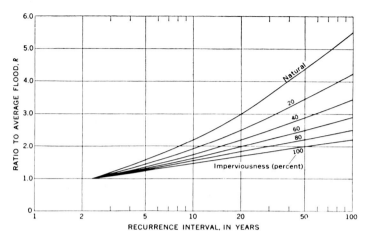

Fig. 9.6 Flood frequency curves for various degrees of imperviousness as ratio of large floods to average annual high-water mark. (From Anderson, 1970.)

politan area. The wide discrepancies in hourly rainfall received can be seen in Table 9.2.

It is startling to note that in both storms a substantial part of the network recorded nothing during the hour when another gauge got its highest amount. During passage of the squall line 65 percent of the area observed no rain during the hour of intense fall elsewhere. During the slow-moving storm the rainfall was more widespread, but even then 25 percent of the area was rain free when the remainder became drenched. It is not uncommon in large communities to experience heavy rain during a thunderstorm in only a small quarter. Large storm systems, while generally not exhibiting quite as intensive rainfalls in very short intervals, have more uniform amounts through a community because of their longer duration. This complicates both modeling and forecasting of urban floods. It also indicates the inadequacy of proceeding on the basis of single rain-gauge observations.

Where multiple rain-gauge observations are available for appreciable intervals of time, probability functions for various parts of a city can be worked out to cover the chances of extreme events for different sectors of a sprawling urban community. This will improve design values for drainage systems. A common practice is also the construction of a hypothetical maximal storm, for a given time interval, say 100 yr, to anticipate potential problems. Here, again, the

TABLE 9.2

Number of Rain Gauges Receiving Specified Amounts of Hourly Rain in Heavy Storms over St. Louis[a]

Number of rain gauges		Rain amounts	
July 9, 1973	July 28, 1974	Inches	Millimeters
146	57	None	None
37	81	0.01–0.10	0.25–2.5
12	15	0.11–0.20	2.6–5.1
7	17	0.21–0.30	5.2–7.6.
3	9	0.31–0.40	7.7–10.2
3	11	0.41–0.50	10.3–12.0
1	5	0.51–0.60	12.8–15.2
2	4	0.61–0.70	15.3–17.8
4	4	0.71–0.80	17.9–20.3
1	2	0.81–0.90	20.4–22.9
1	2	0.91–1.00	23.0–25.4
0	1	1.01–1.10	25.5–27.9
0	1	1.10–1.20	28.0–30.5
6	15	>1.20	>30.5

[a] From Project METROMEX.

use of partial drainage areas is essential. In rolling terrain, or where steep slopes exist in an urban area this procedure is inescapable. Even then surprises can occur in areas where tropical storms may pass occasionally. In those cases past records are usually statistically inadequate to cover what statisticians call an outrider event.

Urbanization certainly aggravates flooding but later problems can now, by using relevant experience, be avoided as the removal of vegetation proceeds and impervious surfaces are increased. Stall *et al.* (1970) have indicated in case of an example (Urbana, Illinois) how characteristics change for four categories of environment from rural to completely urbanized. An excerpt from these studies, shown in Table 9.3, will illustrate how runoff ratios vary, where the runoff factor is defined as $X = R_u/t$, where R_u is the rainfall excess in the urban area and t the duration time. This will result in a maximum runoff $Q_{max} = AX_{yz}$ where A is the area and the quantities y and z are a climatic factor defining habitually and a terrain factor indicating retardations in the runoff, respectively.

It should be noted that these numbers apply to a section of Urbana, Illinois drained by the Boneyard stream, an area of 3.58 mi^2

TABLE 9.3

Urbanization Stages and Their Effects on Stream Discharge in East-Central Illinois[a]

| | | | 2-hr max storm | | | |
| Environment | Impervious ground cover (%) | Recurrence interval (yr) | Rain amount | | Stream discharge | |
			in.	mm	ft³ sec⁻¹	m³ sec⁻¹
Rural	3	2	1.7	432	144	4.1
		10			243	6.9
		50			443	12.5
		100			1145	32.4
⅓ rural, ⅔ urban	25	2	2.1	533	302	8.7
		10			398	16.3
		50			610	17.3
		100			1600	45.3
All urban	50	2	3.15	800	501	14.4
		10			587	16.6
		50			819	23.2
		100			2125	60.2
Intensely urbanized	75	2	3.60	914	632	18.0
		10			746	2.1
		50			928	26.3
		100			2415	68.4

[a] According to Stall et al. (1970).

(9.3 km²). The ratios between the intensely urbanized area to the rural area indicate 8 times greater runoff for the 2-yr maximum rainfall and to nearly 4 times for the projected 100-yr storms. These values can be taken as indicative, but it must again be emphasized that each urban area is governed by its own climatic and topographic conditions.

Natural or man-made lakes in urban areas offer a number of special challenges. Two problems are particularly trying. One is the prevention of eutrophication, which is caused by excessive growth of algae. The other is excessive silting. Eutrophication is usually produced by the discharge of biological wastes into the lakes. This produces not only nutrients for the algal cover but, because of the high oxygen demand of these wastes, robs the lake fauna of an essential element for their survival. Cleanup of domestic and, obviously, of industrial waste waters is a must in urban water management. Silt is harder to control. Every heavy rainfall will carry with it erosion products from urban surfaces and what is left of soil. Abrasion from roofs and streets can be very substantial. In the developing town of Columbia, Maryland, two small lakes, produced by damming up streams, were esthetic and recretional amenities, but heavy rains such as those shown in Fig. 9.2, caused so much silting that dredging is required every few years. Sediment traps can reduce this problem. Control of water quality at the inflows, especially adequate dissolved oxygen is essential. For reservoir-type lakes a large ratio of volume-to-surface area is desirable, i.e., deep reservoirs are desirable for good management (Britton *et al.*, 1975).

References

Anderson, D. G. (1970). Effects of urban development on floods in Northern Virginia. Geological Survey Water-Supply Paper 2001-C, 22 pp. Washington, D. C.

Britton, L. J., Averett, R. C., and Ferreira, R. F. (1975). An introduction to the processes, problems, and management of urban lakes. Geological Survey Circular 601-K, 22 pp. Reston, Virginia.

Gray, D. M. (1973). "Handbook on the Principles of Hydrology." Water Information Center, Inc., Port Washington, New York.

Huff, F. A., and Vogel, J. L. (1977). Maximum minute and hourly rainfalls on METROMEX network, 1971–1975. *Ill. State Water Surv., Circ.* 126, 81 pp.

Landsberg, H. E. (1979). Atmospheric changes in a growing community (The Columbia, Maryland Experience). *Urban Ecology* **4**, 53–81.

Lazaro, T. R. (1979). "Urban Hydrology," 249 pp. Ann Arbor Science Publ., Inc., Ann Arbor, Michigan.

McPherson, M. B. (1979). Urban hydrology. *Rev. Geophys. Space Phys.* **17**, 1289–1297.

Stall, J. B., Terstriep, M.L., and Huff, F. A. (1970). Some effects of urbanization on floods. *ASCE Natl. Water Res. Eng. Mtg., Memphis, Jan. 26–30, 1970*, 29 pp. (Meeting preprint.)

Terstriep, M. L., Vorhees, M. L., and Bender, G. M. (1976). Conventional urbanization and its effect on storm runoff. *Ill. State Water Surv.*, 68 pp. (Contract report.)

10

Special Aspects of Urban Climate

The urban atmosphere has special effects on man and his works. It also affects other biota, especially plants. In this chapter attention will focus on these interactions. It needs little imagination to pinpoint pollutants as the principal agents of these effects. Because of the transdisciplinary aspects of this particular niche of the biosphere its exploration is inadequate and mostly qualitative.

10.1 CORROSION AND DETERIORATION

Monuments, buildings, tombstones, metalwork are silent witnesses of the relentless impact of urban atmospheres on human artifacts. The sad results of urban air pollution on the cultural heritage of mankind are obvious in many old cities. We need only to relate here the fate of the figures supporting the porch roof of the Erechtheum on the Acropolis in Athens, Greece (Fig. 10.1). The pitting and rapid rusting of unpainted parts of motor vehicles in cities is a common experience to Americans. The effects on paints too are uncontested (Akademie der Wissenschaften, 1975).

Many of the urban contaminants combine to cause these effects. However, there is ample observational justification to allocate a major portion of the blame on sulfur dioxide and other strong acids,

Fig. 10.1 Caryatid from the Acropolis in Athens, showing erosion of the sculpture by acid air pollutants (courtesy of Press and Information Office, Embassy of Greece, Washington, D. C.).

their acidification of rainfall and certain halogen (especially chlorine) compounds. The costs of urban deterioration by these pollutants have been estimated in various countries of running into the billions of dollars. Iron, high-carbon steel, zinc, copper, aluminum, nickel, and chromium are among the metals subject to corrosion by pollutants. Among the variety of installations subject to gradual deterioration are power lines, fencing, car parts, and rails.

Vulnerable to the attacks of these pollutants are also paints and textiles, and nylon, among the man-made fibers, is particularly affected. Animal furs and sheep's wool are more resistant. If any of the acids penetrate, even in parts per billion, into buildings they act destructively on papers. There are also reports that books in urban libraries have been adversely affected. The costs of damages to materials are very high. Sulfur compounds and particulates cause annual damage to paint estimated in the neighborhood of a billion dollars. Corrosion of metals is even more costly, estimated at 1.5 billion dollars per year. Oxides of sulfur, oxides of nitrogen, and oxidants destroy textiles and fibers in excess of 0.6 billion dollars per year. They also cause deterioration of rubber and elastomers, also in the neighborhood of 0.5 billion dollars annually.

The most notable effects are noted on building materials. Primary targets for deterioration are limestone, marble, slate, and mortar. The calcium carbonate in these is transformed into sulfate. This is particularly serious in mortar because the change leads to a volume expansion that can cause lifting in the building elements joined by mortar.

Among the most regrettable events is the deterioration of old fresco paintings on the mortared walls of churches and monastaries in old European cities. A typical case concerns the early fourteenth-century frescoes by Giotto in the Scrovegni Chapel at Padua, Italy, which are badly deteriorated. Winkler (1977) has reported the nearly complete destruction of a relief carving, dating back to about 1170, in a church at Opherdicke in West German Westfalia. The calcite in the sandstone used for the relief was nearly completely destroyed. This author also pointed out that the high-temperature contrasts on the walls of insolated buildings and the trapping of rainwater in the building material through leaching leads to deterioration. Some of the same is attributed to the effects of salting the streets in urban areas to melt snow. The attack of the salt through crystallization contributed to the decay of stoney material

TABLE 10.1

Weathering in Urban and Rural Setting[a]

	Effects	
Cause	Rural	Urban
Frost action	Moderate	Extreme
Flaking by heat and moisture	Little	Extensive
Salt efflorescence	Little	Extensive
Solution of carbonates	Little	Very damaging

[a] Adapted from Winkler (1977).

when the salt solution penetrates the cracks. Table 10.1 gives a qualitative comparison of the weathering in urban and rural areas.

The most notable deterioration has been observed in marble. This beautiful white stone has been extensively used since antiquity for monuments, statues, columns, and tombstones. The calcite in this (and other stones) is transformed by acid rain into gypsum, a material which is not only more soluble but also softer and hence more readily eroded by rain. The damage to historical structures such as the Parthenon, Colosseum, and the Taj Mahal is extensive and has brought about attempts to devise countermeasures (Gauri, 1978). But one need not go back to antiquity or the Middle Ages to find evidence of such deterioration. Every urban cemetery has marble tomb stones in it that show various signs of deterioration. In some towns in New England stones in cemeteries have so badly decayed that the inscriptions have become obliterated in less than two centuries.

10.2 URBAN NOISE

The propagation of sound in the atmosphere is governed by the density of the air, which in turn depends on the temperature. Sound waves are subjected to reflection and diffraction. The vertical temperature structure determines the bending of sound rays. For example, a warmer layer of air above a colder layer will bend the waves downward. Yet the simple physical laws are only of limited help in the assessment of sound problems in urban environments.

Scattering by atmospheric turbulence and reverberation from reflecting surfaces in canyon streets force us again into an empirical posture. The wind structure also has a notable influence on sound and its focussing, as well as on shadow zones of diminished sound (Lyon, 1973).

The sources of sound in urban areas are ubiquitous but irregular in intensity and time of occurrence. Their frequency mixture characterize these sounds as noise and the nuisance quality has led to the label "noise pollution." The U. S. Environmental Protection Agency (1973) has stated in that respect that: "So noisy, in fact, is America's urban environment that people living in congested sections of large cities may be hearing far less than they realize; many are developing severe hearing loss."

The sources of noise in urban areas are manifold. Noisy means of transportation are concentrated in or near cities. Aircraft, railroads, and motor vehicles all contribute to the urban din. A number of measurements of the noise levels in urban areas have been made. These are expressed in decibels (dB), a logarithmic scale of the pressure exercised by sound waves. The lowest audible sound is classed as 0 dB, equivalent to a pressure of 0.0002 μbars (2×10^{-9} N m^{-2} = 2 nPa). A sound of 60 dB, generally considered as the beginning of intrusive noise levels, has a million (10^6) times higher pressure than the zero level. And 90 dB is the threshold of potential damage to hearing with a billion higher pressure (2 Pa) than 0 dB. The measurements of sound are so arranged that they comprise the frequencies that are those to which the human auditory system is sensitive (usually labeled as dBA) (Table 10.2).

Traffic obviously contributes most to the noise level and occasionally an ear-piercing motorcycle noise can lift the decibels to 90. Takeoffs at nearby airports will regularly exceed even this level. In some residential areas of cities overflights of airplanes will raise noise to the intrusive level. Undoubtedly the fluctuating quality of urban noise must contribute to the stress it causes. This characteristic of urban noise was very well demonstrated by high-speed records taken on the seventh floor of a building at Columbia University on Amsterdam Avenue in New York City (Ballestin et al., 1970). Short periodic fluctuations of 30 dB in less than a minute are common. These authors found distinct diurnal and weekly variations in noise levels. Weekday noises were lowest from midnight to 4 a.m.; they increase to the 9 a.m. rush-hour level and then stay

TABLE 10.2

Urban Noise Levels[a]

	Noise level (dBA)		
Environment or cause	Min	Mean	Max
Third-floor apartment, next to freeway	75	81	89
Second-floor tenement, New York	62	71	83
Urban shopping center	54	60	71
Urban residential area	47	53	68
near major airport	51	59	93
6 miles from major airport	50	56	70
Suburban residential at outskirts of city	40	45	68
Rural farm	32	39	53

[a] Adapted from Gilluly (1972).

fairly constant into the evening hours. On weekends the noise levels show less diurnal variation but there is more noise in the early morning hours than on weekdays. In the morning and early afternoon hours weekend noise is lower than on weekdays. Data published for Toronto (Koczkur *et al.*, 1973) show a fairly even noise structure through the day; the minimum value in a residential district, but not far from a major thoroughfare, was 41 dB at about 5 a.m.; the noisiest period was during the morning rush hour with a maximum of 82 dB by about 9 a.m.

Traffic noises decrease fairly rapidly with elevation above ground. An experiment on a 39-story building in New York City showed this rather conclusively (Meyerson, 1977). Table 10.3 shows the percent of time a given noise level was exceeded during a midweek 3-hr sampling period in midday. Although this is a very small sample, the general tendency is well represented.

In comparing indoor with outdoor noises in urban residential sectors, a few conclusions can be drawn from limited observations (Bishop and Simpson, 1977). From midnight to about 0700 hr, noise levels indoors are about 15–20 dB below the outdoor level. During that interval outdoor noises influence the noise level indoors. Outdoor levels reach their peaks from noon to 1800 hr, but remain generally above the indoor level. After that indoor levels (presumably because of radio, television, record players, etc.) approach or may even exceed temporarily the subsiding outdoor noise.

TABLE 10.3

Percent of Time Certain Noise Levels Were Exceeded on Four Floors of High-Rise Building

Time exceeded (%)	Noise level (dBA)			
	3rd floor	14th floor	26th floor	37th floor
85	1	<0.5	<0.5	<0.5
80	2	0.5	<0.5	<0.5
75	5	1	<0.5	<0.5
70	14	3	1	<0.5
65	50	13	4	1
60	80	75	60	>90

During the Columbia, Maryland, experiment of atmospheric surveillance in a developing community, some observations of sound levels were made (Landsberg, 1972). These covered the daytime hours from 8:30 (a.m.) to 14:30 (p.m.). Minute by minute observations were made in two settings. One was a small shopping center with a sizeable parking lot, a supermarket and a number of small stores. The other was wooded area in the townsite, but not yet developed. However, a traffic artery was within 400 m. The frequency distributions of noise levels of 360 observations at each site are shown in Fig. 10.2. In the wooded location over 60 percent of the observations were below 40 dB. It was a very quiet environment indeed, with most of the sound produced by the wind rustling the leaves on the trees. A woodpecker caused one of the noisiest interludes. There were a few extraneous traffic sounds from the highway. In contrast, the modal value of noise at the shopping center was between 55 and 60 dB, mostly from traffic, or about 20 dB higher than in the wooded area. During the survey the noise level never sank below 50, and 25 percent of the values were levels above 60 dB. Some values in the 70–80 dB range were from aircraft, others came from emergency vehicles and trash trucks.

Understandably, considerable efforts have been devoted to screening residential sections from traffic noise. This is generally only possible when a reasonable distance between highways or roads and housing exists. In that case shelter belts of trees and shrubbery as well as protective walls can be employed. Numerable observations by Cook and van Haverbeke (1977) have led to the

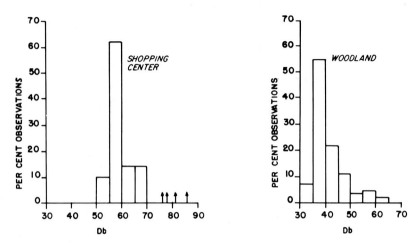

Fig. 10.2 Frequency distribution of various noise levels, for 1-min readings, from 6-hr survey in the Columbia, Maryland, area: left panel in a shopping center; right panel in undeveloped woodland.

conclusion that a combination of solid surfaces with planted strips is best. These authors conclude that a distance of at least 25 m from the center of a roadway to a residence is needed for satisfactory results. One or two rows of dense evergreen trees should be planted as close to the road curb as possible, then a solid barrier and then further trees or shrubbery. Such screening, at distances up to 20 m

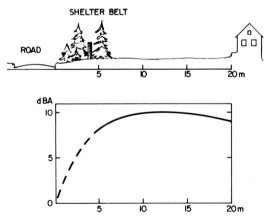

Fig. 10.3 Recommended noise attenuation screen for residence (upper panel) with anticipated noise reduction in dBA (lower panel).

from the belt will result in about 10-dBA noise attenuation over and above the theoretical reduction because of distance from the source. A typical noise screen and attenuation is shown in Fig. 10.3.

Aside from the benefit of such a shelter belt for noise reduction, it also acts as a filter for air pollutants from road traffic, especially aerosols (Neuberger *et al.*, 1967, Rich, 1971). The interception and filter effect of coniferous trees is on an average 34 percent for small particles. Other benefits are the provision of habitats to birds, reduction of runoff, and aesthetic appeal.

10.3 PLANTS IN THE URBAN ATMOSPHERE

For many decades hygienists have urged the preservation and increase of green areas in cities. We read in the classical treatise of Brezina and Schmidt (1937) that parks are the "lungs of the metropolis." The idea here was that the excess carbon dioxide produced in the urban area would be absorbed by the photosynthetic processes of plants and oxygen given off. Although there is a considerable literature on the removal of other contaminants there is little quantitative information on the $CO_2 \rightarrow O_2$ conversion in cities. It is probably a minor contribution to purification of the air. A principal justification for green areas has been to improve the aesthetics of towns (Noyes, 1971) and to provide for playgrounds and other recreational facilities for urban populations.

Yet beyond these highly desirable social aims there is a lively interaction between the urban atmosphere and living plants. In general, the green surfaces mitigate the less-desirable aspects of the urban climate. Within their confines and somewhat beyond, they certainly reduce the stress produced by the heat island. They decrease the noise levels and filter out certain pollutants. They to improve the water balance and reduce runoff. But keeping the plants and trees alive in the urban atmosphere is not easy. First there is the pressure to use the expensive urban land intensively for housing and other buildings. Many have decried the population density which, as Gottmann (1970) described it, has required a "manmade environment in which people are divorced from the natural framework of trees, woods, fields, and so forth." Fortunately, in many cities some of the open areas, such as parks, commons, cemeteries, and even golf courses are protected as public lands. A changed attitude toward urban landscaping is not only protecting these amenities but improving both quality and area of planted surfaces.

Bach and Matthews (1969) have summarized the typical land use of major United States cities. Their findings are shown in Table 10.4.

Much of the urban open space is, of course, quite fragmented in the form of lawns and small gardens in residential sections. Some is represented by shade trees along streets. Although in the aggregate these are highly desirable, they affect only the microclimate wherever they are located. Large, coherent park areas, however, have quite a measurable influence. In warm summer nights the parks are cool, even in the densely built-up area, as Lewis *et al.* (1971) demonstrated so well for Washington, D. C. One of their characteristic surveys is shown in Fig. 10.4, which also shows the route taken. The cooling influence of the large wooded area of Rock Creek Park in a typical traverse near midnight was about 3°C cooler than the average for all points surveyed. The residential area was 0.5°C cooler, the business district about 2°C warmer, and the inner city open spaces close to the average. In daytime, on sunny summer days, the park heats up much more slowly than the built-up section, and even at the time of maximum temperatures it is usually 1–2°C cooler than downtown (Landsberg, 1956).

The beneficial effects of park land in urban areas has been quantitatively established for particulates. Contaminated air moving across parks loses aerosols by impingement on trees and shrubs, and depending on the wind direction, density of tree cover, and width of

TABLE 10.4

Land Use in Some Major United States Cities[a]

Type of land use	Percent of area
Residential	30
Open space	26
Traffic arteries, streets	24
Community facilities (schools, etc.)	8
Industrial	7
Commercial	5

[a] Chicago, Cincinnati, New York, Philadelphia, Washington, D. C. Adapted from Bach and Matthews (1969).

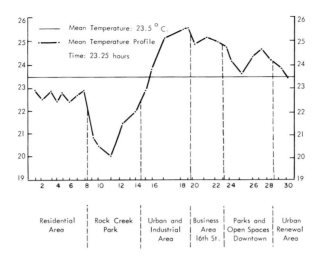

| Residential Area | Rock Creek Park | Urban and Industrial Area | Business Area 16th St. | Parks and Open Spaces Downtown | Urban Renewal Area |

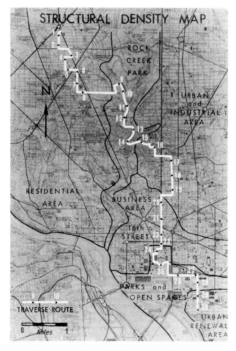

Fig. 10.4 Nocturnal temperature (°C) traverse through Washington, D. C., with various land uses indicated (upper panel). Route and measuring points marked on structural density map (lower panel) (from Lewis *et al.*, 1971).

the planted area, as much as 75% of the coarser fraction of dust (diam. >50 μm) is removed. Other pollutants are also intercepted but the result is often highly damaging to the plants. Acute injuries are produced by sulfur dioxide (SO_2), fluorides, ozone, peroxyacetyl nitrate (PAN), and chlorine compounds (Root and Robinson, 1949). Although all of these are subject to air quality controls because they are also injurious to humans, some plants are sensitive to very small concentrations (Taylor, 1973).

The effect of the urban atmosphere on plant material was first observed last century when Nylander noted in 1866 the lack of lichens in the Luxembourg Garden of Paris (Domrös, 1966), and has been confirmed over the years in many cities. The inverse correlation between pollutant concentrations and frequency of lichen growth on trees in and near cities of the West German Ruhlr region is too high to be coincidental (Domrös, 1966). Although SO_2 has been specifically cited as the cause, this remains controversial. But clearly the cities have become what has been termed "lichen deserts."

Aside from the well-known effects of SO_2, O_3, and fluorides on vegetables and ornamental plants (such as coleus, petunias, and gladiolus), a number of tree species suffer. Ryder (1973) lists the species shown in Table 10.5.

Most of the trees listed in Table 10.5 are more likely to be found in ornamental plantings in home gardens than in public parks. More extensive listings of the sensitivity of tree species to SO_2 and O_3 have been given by Davis and Gerhold (1976). These authors list only seven species as resistant to both of these pollutants, as shown in Table 10.6. This table also has a notation on the salt resistance of

TABLE 10.5

Trees Sensitive to Common Air Pollutants[a]

Tree	Pollutant
Douglas fir	SO_2
Eastern white pine	O_3, SO_2
Larch	SO_2
Norway spruce	SO_2, fluoride
Ponderosa pine	Oxidant, fluoride
Scotch pine	SO_2, fluoride

[a] After Ryder (1973).

TABLE 10.6

Trees Tolerant to SO$_2$ and O$_3$[a]

Species		Salt tolerance[b]
White Fir	*Abies concolor*	(Not given)
Sugar Maple	*Acer sacharum*	(Contradictory evidence)
Western Juniper	*Juniperus occidentalis*	(Not given)
Blue Spruce	*Picca pungens*	Good
Red Oak	*Quercus rubra*	Good
Northern White Cedar	*Thurja occidentalis*	Moderate
Basswood	*Tilia cordata*	Poor

[a] After Davis and Gerhold (1976).
[b] Dirr (1976).

these species, a problem that becomes acute in cities in the higher latitudes where salt is used for snow removal (Dirr, 1976).

There are other stresses on urban plant life besides pollutants. Among these is the illumination problem. In some urban areas the exposure to sunlight is greatly restricted and in many places artificial light changes the natural circadian and annual rhythms of the plants. Another challenge to survival for the ornamental shade trees along sidewalks is lack of water because of inadequate permeable space around the base of the trunk. Efforts to breed plants adapted to these urban adversities are underway. On the other hand, one should not overlook the rather notable advantage offered to some species by the urban heat island. It enables certain trees to survive in a city far beyond its usual latitudinal range. An example are a few magnolias near the center of Boston. Other plants will keep their leaves longer in the fall than in the suburbs and bloom earlier in the spring. But complete symbiosis between plants and urban atmospheres remains to be achieved.

10.4 EFFECTS OF URBAN ATMOSPHERE ON HUMANS

As we have seen, the difference in emitted and suspended pollutants between urban and rural areas is so large that it is not surprising that health effects have been attributed to them. A complete review of these would fill a book, although it has been difficult to pin some cause-and-effect relations down with the rigor demanded by scientific proof (Goldsmith and Friberg, 1977). A distinction must be made here between acute episodes and long-term influences. In the former the relation is clear, even if the specific offending pollutant may not be identified. The slow, insidious health insults are usually only apparent in epidemiological analyses, and these are necessarily vague with regard to the pollutants and their cumulative and delayed impact on human morbidity and mortality. Even though a number of substances can be identified as health-damaging, synergistic effects among these remain to be explored. Current medical opinion on the health effects of regulated pollutants is contained in a critical review by Ferris (1978).

The pathways through which the pollutants enter the body are manifold but most of them are probably absorbed through the respiratory system. Some are ingested and some affect eyes and skin. The fact that inhaling is the principal mechanism of human exposure to air pollutants complicates matters further because of the large number of smokers in the population and the unknown interactions of tobacco smoke and pollutants. In many instances the pollutants provoke allergic reactions and there is clear evidence that great differences exist in the instantaneous physiological responses of individuals against pollutants at concentrations common in urban areas.

For some of the catastrophic air pollution episodes, we know the beginning and the end. Under extreme stagnation conditions, with low-level inversions and calm air, pollutants accumulated to high concentrations, often with fog formation. Cases of acute respiratory disease and deaths followed. In most instances it is not known whether or not a specific pollutant was the imediate cause of these serious reactions. Meteorologically it is well known that under such circumstances the changes of all pollutants emitted in an urban area show a high correlation. In the most notorious pollution events in the last half-century it has been made plausible that SO_2 is to be most suspected. Table 10.7 lists a few of the major catastrophic urban pollution events in the past half-century.

These events greatly contributed to legislation to remedy the situation by emission controls. The London episode is depicted in Fig. 10.5, together with the simultaneous measurements of SO_2. This probably formed sulfuric acid in the coexistent fog or in people's

TABLE 10.7

Major Air Pollution Episodes

Locality	Date	Number of deaths	Number sick
Liège, Meuse Valley	Dec. 1930	60	6000
Donora, Pennsylvania	Oct. 1948	20	5900
London, England	Dec. 1952	4000	?
New Orleans	Oct. 1953	2	200
New York	Nov. 1953	165	?

respiratory system. The 4000 excess deaths listed in Table 10.7 are most likely a conservative estimate. It is notable from the graph that a high death rate continued to prevail even after the SO_2 values subsided to tolerable levels. In all these episodes it was the older persons and babies that suffered most.

Age-adjusted morbidity rates for Japanese cities showed four to five times the incidence of respiratory disease, as shown in Fig. 10.6 (*Natl. Air Pollution Control Administration*, 1970). The Public

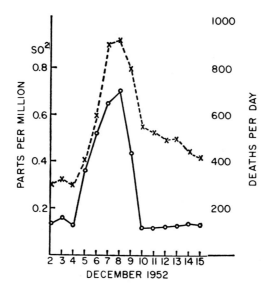

Fig. 10.5 Mortality in London, England (left scale), during air pollution episode of December 1952 (dashed line). The pollution level is characterized by the simultaneous SO_2 concentration (solid line, left scale).

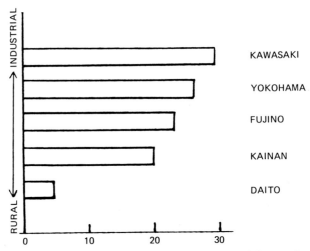

Fig. 10.6 Age-adjusted morbidity rate (cases per 1000) for respiratory ailment, represented as a profile from an industrial-urban area to a rural sector in Japan (from National Air Pollution Control Administration, 1970).

Health Service also found a linear relation between the sulfate content of air and the yearly number of cases of respiratory disease lasting more than 7 days in women for a number of United States cities (Fig. 10.7). Qualitatively, the relative number of persons affected by chronic bronchitis and emphysema, even among non-smokers is significantly higher in cities than in the country, and so is the relative incidence of asthma attacks. Unfortunately, there are so many variables, both in the atmospheric concentration and composition of the pollutants and in the responding population, that it is difficult to be very specific with regard to the health effects of urban air. This is particularly true for pollutants that are slow acting and that may not produce acute but rather chronic manifestation. This applies particularly to mortality. Although the effects are well demonstrated for the several episodes, evidence for effects on death rates by low-level persistent pollution is shaky (Schimmel and Greenburg, 1972).

Only for cadmium has there been a claim that its concentration may have an influence on mortality from hypertension and arteriosclerotic heat disease (Carroll, 1966). A highly significant correlation coefficient of 0.76 was found between atmospheric cadmium concentrations in 28 United States cities and mortality from heart dis-

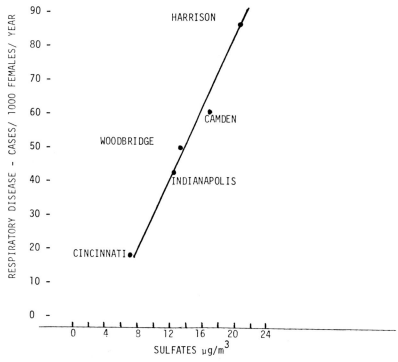

Fig. 10.7 Respiratory disease rate per 1000 women for cases lasting longer than a week as related to atmospheric sulfate concentrations in $\mu g\ m^{-3}$ in various U. S. cities (from National Air Pollution Control Administration, 1970).

ease (excluding rheumatic heart disease). The causality chain of this connection remains unknown (Fig. 10.8).

Table 10.8 lists a number of polluting substances in urban air, their typical concentration, and their presumed health effects. The table gives only a small glimpse at the potentially health-related air pollutants. In all urban atmospheres innumerable chemical reactions take place (Tuesday, 1971). Many otherwise innocent elements participate as catalysts in this process, i.e., iron and vanadium, which are common in the urban particulate suspensions. The toxicity of many of the new compounds in the prevalent concentrations has not yet been determined.

Many of the substances with well-established adverse health effects in urban air are from the effluents of motor vehicles. This applies particularly to lead. One can trace this element in suspended

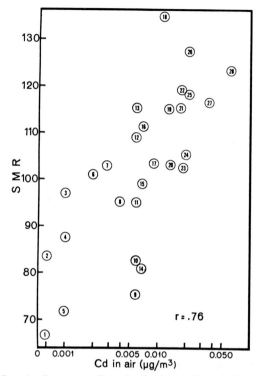

Fig. 10.8 Standardized mortality rates for heart diseases (except rheumatic type), 1959–1961, as related to atmospheric cadmium concentrations in μg m^{-3}, at various United States localities: (1. Las Vegas, NV, 2. Eugene, OR, 3. Medford, OR, 4. Chattanooga, TN, 5. Albuquerque, NM, 6. Omaha, NE, 7. Gary, IN, 8. Los Angeles, CA, 9. Oklahoma City, OK, 10. Phoenix, AZ, 11. Akron, OH, 12. Racine, WI, 13. Wilmington, DE, 14. Tucson, AZ, 15. Youngstown, OH, 16. Cincinnati, OH, 17. Canton, OH, 18. Scranton, PA, 19. New York, NY, 20. Columbus, OH, 21. Charleston, WV, 22. Newark, NJ, 23. Indianapolis, IN, 24. Waterbury, CT, 25. Bethlehem, PA, 26. Philadelphia, PA, 27. Allentown, PA, 28. Chicago, IL (Carroll, 1966). (Reproduced by permission, copyright American Medical Association.)

and deposited dust as it decreases with distance from major traffic arteries. Its concentration in human blood and organs also follows this decrease with respect to location of residences and workplaces. The health effects are particularly notable in small children, who are exposed to this hazard by playing in the streets. The hyperactivity syndrome has been attributed to this hazard. The gradual elimina-

TABLE 10.8

Common Urban Air Pollutants Suspected of Causing Health Problems[a]

Substance	Typical concentration (m^{-3})	Health effect
Sulfur dioxide	10–30 μg	Respiratory diseases
Carbon monoxide	5–25 mg	Cardiovascular and nervous-system symptoms
Ozone and oxidant	50–150 μg	Eye and respiratory irritation
Nitrogen dioxide	100–200 μg	Respiratory irritation
Lead	0.5–2.5 μg	Nervous-system symptoms in children
Cadmium	30–80 ng	Heart and kidney disorders
Polycyclic hydrocarbons	1–5 ng	Lung cancer
Asbestos	1–5 ng	Lung cancer

[a] Sources: National Academy of Sciences (1972), Stern *et al.* (1973), Weselowski *et al.* (1977), Zoller *et al.* (1977), Colucci (1976), Goldsmith and Friberg (1977), and Kneip *et al.* (1979).

tion of leaded fuels will eventually abate the high lead concentrations in urban air.

Carbon monoxide also diminishes with distance from areas of dense motorized traffic. Even nonsmokers living on lower floors in dwellings close to congested urban thoroughfares can show as much as 5 percent of their hemoglobin transformed into carboxyhemoglobin (COHb). This is an amount already quite worrisome for persons with impaired heart function. At greater distances and in upper floors the concentrations penetrating indoors from the outside air are rarely high enough to cause concern. There are, however, many circumstances when city dwellers are exposed to carbon-monoxide concentrations far in excess of permissible levels, such as in tunnels and parking garages. Persons exposed to these conditions can experience headaches. Individuals travelling in cars in slow-moving heavy traffic are about 10 percent of the time exposed to CO values in excess of 35 mg m^{-3} (Brice and Roesler, 1966). On lengthy trips in school buses, children have been exposed to 15–20 mg m^{-3}. These conditions may at least contribute to drowsiness.

According to Ferris (1978), 20 ppm (24 mg m^{-3}) of CO leads to 0.8 percent COHb after 1-hr exposure and 2.8 percent after an 8-hr exposure; for 50 ppm (59 mg m^{-3}) CO the COHb level would be 2.5 percent after 1 hr and 7.5 percent after 8 hr of exposure. At levels of about 2.5 percent of COHb persons with certain cardiovascular diseases (*angina pectoris*, e.g.) may show some trouble; 4–5 percent will lead to headaches and lassitude. Higher values spell real health troubles.

Penetration into dwellings depends on the building materials, circulation conditions, and opening of windows. Wherever there is rapid air intake from outdoors to indoors as much as ⅔ the outdoor value has been measured inside (Thompson *et al.*, 1973). It must be realized, however, that there are CO sources indoors, such as gas heaters, fireplaces, and other combustion devices. Cigarette smoking can increase CO concentrations indoors considerably above the ambient outdoor values, but other gaseous pollutants usually remain a fraction of the outdoor values (Benson *et al.*, 1972). There is generally a time lag between the outdoor and indoor variations. Both concentration and lag depend decisively on the construction method. The lag and some smoothing of fluctuations is shown in Figure 10.9, taken from a Japanese example. It reflects the particulate concentra-

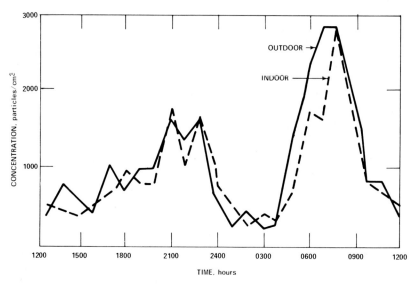

Fig. 10.9 Diurnal variation of indoor and outdoor concentration of particles in an apartment in Toyonaka City, Japan (May 21–22, 1956) (Benson *et al.*, 1972).

tion in a typical Japanese structure. In United States buildings the values usually run between 25 and 75 percent of the outdoor concentrations.

Schaefer *et al.* (1972) related the dust deposited indoors to the outdoor weight of particulate matter per unit volume of air. The correlation coefficient at homes in Chicago was 0.7. In a brief review Georgii (1973) relates that in relatively clean air indoor SO_2 concentrations are 80 percent of the outdoor values but for outdoor values of SO_2 over 1 μg m^{-3} indoors still 30–40 percent prevail. For small CO values of about 12 μg m^{-3}, in the absence of smoking or gas appliances, 60 percent of the outdoor value were noted.

The urban heat island during summer must also be blamed for adverse health effects. Heat stress depends on a large number of variables both in the environment and in the reacting human individuals. It depends on the metabolic rate of individuals, ability to perspire, weight, and age. The environmental conditions of concern are temperature, humidity, radiation load, and wind speed. In some instances clothing is also of considerable influence. A number of studies have related temperature conditions alone to the urban summer mortality rate. Even this primitive approach has yielded quite unequivocal results. Periods of high temperature have shown mortality far in excess of values expected by statistical experience.

The summer of 1966 brought several weeks of excessive heat to the central and eastern United States. The high death rates became immediately obvious to the health authorities in New York City and St. Louis. Figure 10.10 depicts the weekly death rates for the summer of 1966 in New York City and the corresponding temperatures. Deaths exceeded the expected by 50 percent (Landsberg, 1969). Clearly, such high death rates, corresponding to an infectious epidemic, led to closer inspection of the causes and analysis of which segments of the population had suffered most. It was immediately noted that there were a large number of deaths during the period from strokes. In fact, many states from the Great Plains to the Atlantic Seaboard had in July 1966 over 20 percent excess deaths from stroke (Helfand and Bridges, 1971). On the third of the month, the beginning of the heat wave, the LaGuardia Airport weather station in New York City reported its all-time record: maximum temperature of 41.7°C (107°F), far above the normal body temperature of 36.5°C (98°F).

A careful epidemiological study (Schuman, 1972) indicated that

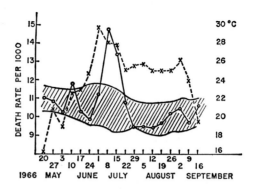

Fig. 10.10 Weekly death rates (solid line, left scale) in New York during summer of 1966. Dashed line (right scale) represents weekly mean temperature. Hatched band represents 95 percent confidence limits of expected death rates.

while cardiovascular accidents (CVA), which include strokes, were considerably in excess of expectation, arteriosclerotic heart disease, hypertension, diabetes, and respiratory diseases contributed heavily to the excess. It also became immediately obvious that the middle aged between 45 and 64 yr and the older groups were particularly at risk. Those over 80 died in very large numbers (Fig. 10.11). A rise was noted even among infants.

In St. Louis, cardiovascular accidents showed a distinct peak in July 1966. During the hottest period, deaths from heat stroke (overheating) were very notable. About 56 percent more persons died in that city, where temperatures exceeded 32°C (90°F) on 24 days. In both New York and St. Louis the women were twice as heavily afflicted as men. No information on obesity was available, although it is well known that overweight persons are particularly at risk in hot environments. Remarkable also was the fact that there were considerable differences in the death rates in various city areas. Sectors where the poor and older residents lived had much higher rates than some of the areas inhabited by more affluent and younger people. Not only does that reflect a difference in air-conditioned dwellings, but as we noted in the sections dealing with the radiation climate and the heat island, the densely settled tenement districts in cities tend to have higher temperatures than the less-intensely developed quarters.

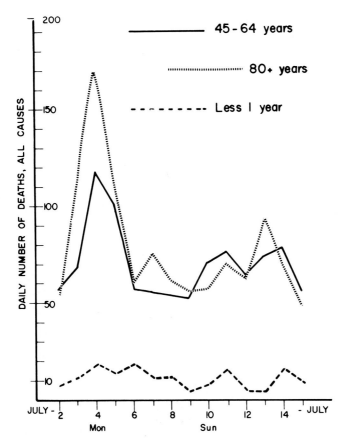

Fig. 10.11   Daily number of deaths from all causes in 1966 heat wave in New York City by age groups (from Schuman, 1972).

It might be noted here in passing that in United States cities there is often a notable bulge upward in deaths in the early part of the summer, when the first spell of high temperatures occurs. This might be ascribed to overheating from unsuitable clothing. People are still wearing warm clothing from the cool spring period and also air-conditioning units are not yet functioning.

Many years show unfortunately similar experiences in one or several cities in the United States. Usually heat waves cover fairly substantial areas so that, especially in the East a number of cities are affected. A notable case occurred in 1975. On July 31, 1975 a fairly normal number (for the season) of deaths in the country was 4934. The next day a heat wave hit the Northeast and the deaths soared to 5733 and for 4 days about 2000 excess deaths occurred. This sudden increase is compared to Philadelphia daily maximum temperatures (Hodge, 1978).

The notorious heat wave of 1980 when temperatures in cities like Dallas, Texas, exceeded 38°C (100°F) for weeks on end led to over 1200 deaths by hyperpyrexia (heat stroke).

Heat deaths are not a experience unique to the United States, the bioclimatically conscious city of Vienna, Austria, had investigated such cases. A notorious hot period there was in July 1957. There were many calls to rescue squads for heat-related incidents (Topitz, 1964). On July 6 air temperature reached 36.4°C (97.5°F). There were 28 calls to the rescue system for heat collapse. On the next day the maximum temperature reached 37.3°C (99.1°F). There were 20 calls for heat collapse and 106 deaths (86 percent above normal). On the third day the thermometer reached 38.3°C (100.9°F). There were 60 heat-collapse cases and 123 deaths (116 percent above normal).

From statistics on heat-syndrome incidence in New Orleans (U. S. Department of Health, Education, and Welfare, 1965), a city with subtropical climatic conditions, it was possible derive a probabilistic relation to the daily maximum temperature (Landsberg, 1973). Figure 10.12 shows that below 27.5°C less than 1 percent of the heat collapses occurred, but nearly 80 percent with temperatures above 30°C. In that area the commonly prevalent high humidities undoubtedly played a role. Physical exertion usually contributes to heat exhaustion but it is also significant that upon entering hospitals many of the afflicted showed evidence of myocardial disease in their electrocardiograms.

The temperature, as measured at a meteorological station, often in an entirely different environment than the inner city, is a very deficient corollary to urban health effects. In New York City no one would claim that the Central Park Observatory or La Guardia Airport are representative of the real bioclimate of business and tenement sections. One element particularly lacking is the heat load imposed by radiative temperatures surfaces, such as pavements and walls. Clarke and Bach (1971) presented clear-cut evidence for this heat load. They showed that the heat fluxes from pavement impinging on persons are nearly 50 percent higher than over nearby

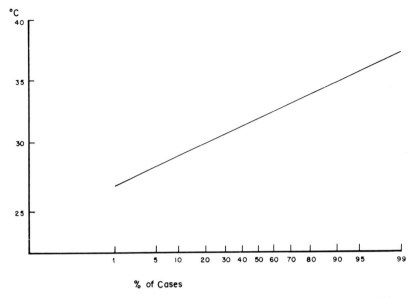

Fig. 10.12 Cumulative heat stroke probabilities in New Orleans as related to daily maximum temperatures (from Landsberg, 1973).

grass surfaces. This can greatly add to the burden, in addition to a high-air temperature. Although a theoretical framework for radiative fluxes in various latitudes and seasons has been developed (Terjung and Louie, 1971), actual measurements, dealing with persons in the streets, their specific clothing, and their physiological reactions are lacking. Chamber experiments are likely to be inadequate.

There have been some attempts to model the human being, metabolizing at various rates, in the urban environment. The radiant exchange with surfaces and walls at various temperature levels has been calculated (Plumley, 1977). Although this is entirely formalistic, the calculated values give some measure of this heat load. In Table 10.9 it is stated in terms of the upper limit of comfort for a person in an ordinary business suit (1 clo unit).[1] If the temperature given is exceeded, an uncomfortable radiant heat load is to to be expected.

From wind-tunnel studies, estimates have also been made of wind

---

[1] clo is the clothing insulation that will permit the passage of 1 cal m^{-2} hr^{-1} with a temperature gradient of 0.18°C between two surfaces: 1 clo 0.18°C/(cal m^{-2} hr^{-1}).

TABLE 10.9

Upper Limits of Radiant Temperatures (°C) for Comfort Conditions

	Activity		
Surface	Sitting	Walking slowly	Walking briskly
Person (in center of 4 × 4-m square)	35	32	23
Wall (at 1 m distance)	35	23	19

effects on comfort conditions in streets and on corners where wind is frequently channeled and broken up into eddies (Arens and Ballanti, 1977). One can easily show that winds may be low and tolerable on one side of the street and highly disagreeable on the other. This will, of course, in the cold season notably affect the wind-chill factor.

Some sporadic measurements on the overall energy fluxes as they impinge on people have been made by Stark and Miller (1977), but all they show is the extraordinary variability in time and urban space.

Although most of the existing literature on heat effects in urban areas have used only maximum temperatures, there are better measures. Ideally observations of the wet-bulb globe thermometer ought to be used because that instrument integrates temperature, humidity, radiation, and wind effects. No systematic observations with this equipment in urban environments are available. A substitute, using two of the elements, is an improvement over the use of temperature alone. This makes use of dry- and wet-bulb temperature measurements, which are fairly commonly available. These two quantities determine the so-called effective temperature (ET), a designation that indicates an arbitrary index, devised by air-conditioning engineers, representing the sensation of warmth by the human body. It is given as a temperature for still, water-vapor saturated air (100 percent relative humidity) that would induce the same sensation as the prevalent ambient air. This has been found to be a physiologically reasonable index, even though there are wide divergences of individual reactions (Sohar et al., 1978).

It has been shown (Thom, 1959) that the ET over the major portion of its range under outdoor conditions can be approximated by

$$ET = 0.4(T_a + T_w + 4.8) \tag{10.1}$$

where T_a is the air (dry-bulb) temperature and T_w the wet-bulb temperatures.

This quantity is also often referred to in the literature as *temperature-humidity index* (THI) or *discomfort index* (DI). This correlates well with heat sensations with a value of 20° ET set as "comfortable." Values above 25° ET fall into the range where heat collapse is possible. Tout (1978) discussed the effect of high ETs during a 1976 hot spell in parts of the British Isles. There again increased numbers of deaths were noted when the ET stayed in daytime above 20° for a period from the last week in June through the first ten days in July. London weekly deaths rose from 1475 during the week ending June 25 to 1956 in the week ending July 2. Among the age group of more than 65 yr old the number of deaths rose from 1050 in the earlier interval to 1483 in the later, an increase of 41 percent. On June 26 and 27, 1500 hr, ET was above 25°, clearly in the danger zone.

Other attempts have been made to relate urban summer mortality in a systematic fashion to a discomfort measure (Hauleitner, 1977). For this purpose a measure of the maximal value of the DI proved to be of some value. This modified DI_s (the subscript standing for summer) is defined as

$$DI_s = 0.55T_{max} + 0.2T_p + 5.2 \tag{10.2}$$

where T_{max} is the mean annual maximum temperature 1°C and T_p the mean dew point at 1400 hr at the particular locality.

The 25 localities investigated by Hauleitner (1977) ranged from 2000 to 1.6 million inhabitants. The yearly death rates, expressed in percent of deaths of population in the particular town, showed a fairly good linear correlation with DI, as one can see from Fig. 10.13. The published data do not permit an analysis for summer deaths only and for specific age classes. Also such other variables as medical facilities were not included in the study.

One other biometeorological factor in the urban milieu is also related to discomfort. High levels of temperature and humidity prevent people from getting rid of their metabolic heat. They are aware of this only subconsciously. It makes them irritable; it leads to behavior problems in nonairconditioned schools and often leads to violence in the streets. Especially in the evenings and nights when urban tenement apartments remain too hot for sleeping, residents take to the streets for relief. In the New York heat wave of July 1966

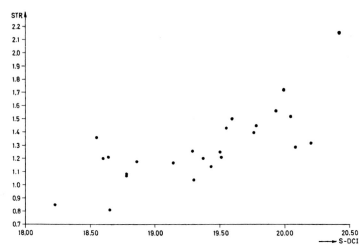

Fig. 10.13 Death rates (ordinate) as related to summer discomfort index in Austrian towns (from Hauleitner, 1977).

homicides rose more than any other cause of deaths, 138.5 percent of the value to be expected for a normal period (Schuman, 1972).

When ETs rise rapidly in urban environments there is evidence that minor incidents or provocations can lead to civil disorders. A pertinent case are the August 1965 disorders in Los Angeles. As is shown in Fig. 10.14 the evening ETs rose rapidly between the 8th and 10th of that month and when they increased above the 25° limit the disturbances started (Landsberg, 1973).

Only limited information is available for potential urban effects in the cold season. A few observations on a clear winter night in Christchurch, New Zealand, showed that rural areas imposed a 25 percent greater radiant energy loss than the business district. This reduced the clothing needs from 3 clo in the rural area to about 2.5 clo in the city (Tuller, 1980). Other studies have shown changes in death rates, inversely related to temperatures in winter. But there is no evidence that urban climate per se had any influence because no comparisons with rural areas were made (Momiyama and Katazawa, 1972).

Some reports also refer to a greater contamination of urban air by pathogenic microorganisms than rural air. The presence of intestinal bacilli points to a relation to population concentrations (Grigorash-chenko, 1963).

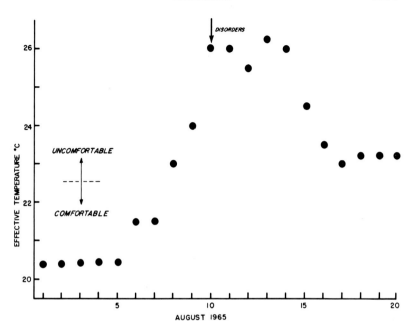

Fig. 10.14 Outbreak of civil disorders in Los Angeles in August 1965 as related to rapid rise of effective temperatures (ET) (from Landsberg, 1973).

The field of urban biometeorology remains a largely unexplored territory that should attract more attention in the future.

References

Akademie der Wissenschaften (1975). Schwefeloxide in der Atmosphäre, Bundesministerium für Gesundheit und Umweltschutz, Wien (Austria), 121 pp.

Arens, E., and Ballanti, D. (1977). Outdoor comfort of pedestrians in cities. *Proc. Conf. Metropolitan Physical Environ., USDA Forest Serv. Tech. Rept.* NE-25, pp. 115–129. Upper Darby, Pennsylvania.

Bach, W., and Matthews, E. (1969). The importance of green areas in urban planning, Paper presented at workshop: *Bioclimatology and Environmental Health, Cincinnati, July 14–16, 1969,* 20 pp. U. S. Dept. of Health, Education and Welfare, Public Health Service, Washington, D. C.

Ballestin, H., Bauman, S., Morganti, S., Pressman, N., Rando, R., and Shaper, R. (1970). Urban noise pollution. *Columbia Eng. Q.,* Feb. 1970, 26–28.

Benson, F. B., Henderson, J. J., and Caldwell, D. E. (1972). Indoor–outdoor air pollution relationships: A literature review. *U. S. Environ. Prot. Agency, Publ.* No. AP-112, 75 pp. Research Triangle Park, North Carolina.

Bishop, D. E., and Simpson, M. A. (1977). Outside and inside noise exposure in urban and suburban areas. *Proc. Conf. Metropolitan Physical Environ., USDA Forest Serv. Tech. Rept.* NE-25, pp. 183–194. Upper Darby, Pennsylvania.

Brezina, E., and Schmidt, W. (1937). "Das künstliche Klima in der Umgebung des Menschen," pp. 196–199. Ferdinand Enke Verlag, Stuttgart.

Brice, R. M., and Roesler, J. F. (1966). The exposure to carbon monoxide of occupants of vehicles noving in heavy traffic. *J. Air Pollut. Control Assoc.* **16,** 597–600.

Carroll, R. E. (1966). The relationship of cadmium in the air to cardiovascular disease death rates. *J. Am. Medical Assoc.* **198,** 267–269.

Clarke, J. F. and Bach, W. (1971). Comparison of the comfort conditions in different urban and suburban microenvironments. *Int. J. Biometeorol.* **15,** 41–54.

Colucci, A. V. (1976). Sulfur Oxides: Current status of knowledge. *Electric Power Res. Inst. Rept.* EA-316, Palo Alto, California.

Cook, D. I., and van Haverbeke, F. (1977). Suburban noise control with plant materials and solid barriers. *Univ. of Nebraska, Res. Bull.* EM 100, 74 pp. Lincoln, Nebraska.

Davis, D. D., and Gerhold, H. D. (1976). Selection of trees for tolerance of air pollutants. *In* "Better Trees for Metropolitan Landscapes," (Santamour, Gerhold, and Little, eds.), *USDA Forest Serv. Tech. Rept.* NE-22, pp. 61–75. U. S. Govt. Printing Off., Washington, D. C.

Dirr, M. A. (1976). Salts and woody-plant interactions in the urban environment. *In* "Better Trees for Metropolitan Landscapes," (Santamour, Gerhold, and Little, eds.), *USDA Forest Serv. Tech. Rept.* NE-22, pp. 103–111. U. S. Govt. Printing Off., Washington, D. C.

Domrös, M. (1966). Luftverunreinigung und Stadtklima im rheinisch-westfälischen Industriegebiet und ihre Auswirkung auf den Flechtenbewuchs der Bäume. *Arb. Rheinischen Landeskunde,* H. 23, 132 pp. Ferd. Dümmler Verlag, Bonn.

Ferris, B. G., Jr. (1978). Health effects of exposure to low levels of regulated air pollutants (a critical review). *J. Air Pollut. Control Assoc.* **28,** 482–497.

Gauri, K. L. (1978). The preservation of stone. *Sci. Am.* **238** (6), 126–136.

Georgii, H. W. (1973). Über das Eindringen von Luftverunreinigungen in Gehäude. *Hippokrates* **44,** 327–329.

Gilluly, R. H. (1972). Noise: The unseen pollution. *Sci. News* **101,** 189–191.

Goldsmith, J. R., and Friberg, L. T. (1977). Effects of Air Pollution on Human Health. *In* "Air Pollution," (A. C. Stern, ed.), 3rd ed., Vol. II, pp. 457–610. Academic Press, New York.

Gottmann, J. (1970). The green areas of Megalopolis. *In* "Challenge for Survival," (P. Dansereau, ed.), pp. 61–65. Columbia Univ. Press, New York.

Grigorashchenko, O. Iu. (1963). Data on the study of bacterial contamination of atmospheric air of the city of Odessa (translated title). *Mikrobiol. Zh.* **25,** 35–41.

Hauleitner, F. (1977). Die Abhängigkeit der Sterblichkeit vom örtlichen Temperatur-Feuchte-Milieu. *Wetter Leben* **29,** 26–34.

Helfand, L. A., and Bridges, C. (1971). Hippocrates, thermal stress, and stroke mortality—1966. *Weatherwise* **24**, 100–104.

Hodge, W. (1978). Weather and mortality, EDIS, Sept. 1978, pp. 12–14. Natl. Oceanic and Atmospheric Administration, Rockville, Maryland.

Kneip, T. J., Mukai, F., and Daisey, J. M. (1979). Trace organic compounds in the New York City atmosphere, Pt. 1. Electric Power. Res. Inst. Rept. No. EA-1121, Palo Alto, California.

Koczkur, E., Broger, E. D., Henderson, V. L., and Lightstone, A. D. (1973). Noise monitoring and a sociological survey in the city of Toronto. *J. Air Pollut. Control Assoc.* **23**, 105–109.

Landsberg, H. E. (1956). Nature's air conditioner. *Am. For.* **62** (8), 17 and 63.

Landsberg, H. E. (1969). "Weather and Health," 148 pp. Doubleday, Garden City, N. Y.

Landsberg, H. E. (1972). Noise increase in an urbanizing area. *J. Wash. Acad. Sci.* **62**, 329–331.

Landsberg, H. E. (1973). Climate of the urban biosphere. *In* "Biometeorology" (S. W. Tromp, W. H. Weihe, and J. J. Bouma, eds.), Vol. 5, Pt. II, pp. 71–83.

Lewis, J. E., Jr., Nicholas, F. W., Seales, S. M., and Woollum, C. A. (1971). Some effects of urban morphology on street level temperatures at Washington, D. C., *J. Wash. Acad. Sci.* **61**, 258–265.

Lyon, R. H. (1973). Propagation of environmental noise. *Science* **179**, 1083–1090.

Meyerson, N. L. (1977). Study of traffic noise levels at various heights of a 39-story building. *Proc. Conf. Metropolitan Physical Environ., USDA Forest Serv. Tech. Rept.* NE-25, pp. 195–201. Upper Darby, Pennsylvania.

Momiyama, M., and Katayama, K. (1972). Deseasonalization of mortality in in the world. *Int. J. Biometeorol.* **16**, 329–342.

National Academy of Sciences (1972). "Lead (Airborne Lead in Perspective)," 329 pp. Washington, D. C.

National Air Pollution Control Administration (1970). Air quality criteria for sulfur oxides. *NAPCA Publ.* No. AP-50. U. S. Dept. of Health, Education and Welfare, Washington, D. C.

Neuberger, H., Hosler, C. L., and Kocmond, W. C. (1967). *In* "Biometeorology," (S. W. Tromp and W. H. Weihe, eds.), Vol. 2, Pt. 2, pp. 693–702. Pergamon Press, Oxford.

Noyes, J. H. (1971). Managing trees and woodlands to improve the aesthetics of communities. *In* "Trees and Forests in an Urbanizing Environment," Univ. of Massachusetts, Holdsworth Natural Resources Center, Planning & Resource Development Series No. 17, pp. 115–120. Amherst, MA.

Plumley, H. J. (1977). Design outdoor urban spaces for thermal comfort. *Proc. Conf. Metropolitan Physical Environ., USDA Forest Serv. Tech. Rept.* NE-25, pp. 152–162. Upper Darby, Pennsylvania.

Rich, S. (1971). Effects of trees and forests in reducing air pollution. *In* "Trees and Forests in an Urbanizing Environment," Univ. of Massachusetts, Holzworth, Natural Resources Center, Planning and Resources Department Series No. 17, pp. 29–33.

Root, I. C., and Robinson, C. C. (1949). City trees. *In* "The Yearbook of Agriculture 1949 (Trees)," pp. 43–48. USDA, Govt. Printing Office, Washington, D. C.

Ryder, E. J. (1973). Selecting and breeding plants for increased resistance to air pollutants. *In* "Air Pollution Damage to Vegetation," (J. A. Naegele, ed.), *Advan. Chem. Ser.* **122**, pp. 75–100. Am. Chem. Soc., Washington, D. C.

Schaefer, V. J., Mohnen, V. A., and Veirs, V. R. (1972). Air quality of American homes. *Science* **175**, 173–175.

Schimmel, H., and Greenburg, L. (1972). A study of the relation of pollution to mortality (New York City, 1963–1968). *J. Air Pollut. Control. Assoc.* **22**, 607–616.

Schuman, S. H. (1972). Patterns of urban heat-wave deaths and implications for prevention: Data from New York and St. Louis during July, 1966. *Environ. Res.* **5**, 59–75.

Sohar, E., Birenfeld, C., Shoenfeld, Y., and Shapiro, Y. (1978). Description and forecast of summer climate in physiologically significant terms, *Int. J. Biometeorol.* **22**, 75–81.

Stark, T. F., and Miller, D. R. (1977). Effect of synthetic surfaces and vegetation in urban areas on human energy balance and comfort. *Proc. Conf. Metropolitan Physical Environ., USDA Forest Serv. Tech. Rept.* NE-25, pp. 139–162. Upper Darby, Pennsylvania.

Stern, A. C., Wohlers, H. C., Boubel, R. W., and Lowry, W. P. (1973). "Fundamentals of Air Pollution," 492 pp. Academic Press, New York.

Taylor, O. C. (1973). Acute responses of plants to aerial pollutants. *In* "Air Pollution Damage to Plants" (J. A. Naegele, ed.), *Advan. Chem. Ser.* **122**, 9–20. Am. Chem. Soc., Washington, D. C.

Terjung, W. H., and Louie, S. S-F. (1971). Potential solar radiation climates of man. *Ann. Assoc. Am. Geogr.* **61**, 481–500.

Thom, E. C. (1959). Discomfort index. *Weatherwise* **12**, 57–60.

Thompson, C. R., Hensel, E. G., and Kats, G. (1973). Outdoor–indoor levels of six air pollutants. *J. Air Pollut. Control Assoc.* **23**, 881–886.

Topitz, A. (1964). Hitzekollaps und Witterung. *Wetter Leben* **16**, 89–104.

Tout, D. G. (1978). Mortality in the June–July 1976 hot spell. *Weather* **33**, 221–226.

Tuesday, C. S. (1971). "Chemical Reactions in Urban Atmospheres," 287 pp. American Elsevier, New York.

Tuller, E. S. (1980). Effects of a moderate-sized city on human thermal bioclimate during clear winter nights. *Int. J. Biometeorol.* **24**, 97–106.

U. S. Department of Health, Education, and Welfare (1965); Heat syndrome data from selected hospital record survey. Report, Contract OCD-OS-62-100 subtask 1221A Public Health Service, Washington, D. C., 82 pp.

U. S. Environmental Protection Agency (1972). Noise pollution, 13 pp. Superintendent of Documents, Washington, D. C.

Weselowski, J. J., John, W., and Kaifer, R. (1977). Lead source identification by multielement analysis of diurnal samples of ambient air. *In* "Trace Elements in the Environment" (E. L. Kothny, ed.), *Advan. Chem. Ser.* **123**, 1–16. Am. Chem. Soc., Washington, D. C.

Winkler, E. M. (1977). Stone decay in urban atmospheres. *Eng. Geol. Case Histories,* No. 11, pp. 53–58. Geological Society of America.

Zoller, W. H., Gordon, G. E., Gladney, E. S., and Jones, A. G. (1977). The sources and distribution of Vanadium in the atmosphere. *In* "Trace Elements in the Environment" (E. L. Kothny, ed.), *Advan. Chem. Ser.* **123**, 31–47. Am. Chem. Soc., Washington, D. C.

11
Urban Planning

The knowledge we have acquired about urban climates should not remain an academic exercise on an interesting aspect of the atmospheric boundary layer. It should be applied to the design of new towns or the reconstruction of old ones. The purpose is, of course, to mitigate or eliminate the undesirable climatic modifications brought about by urbanization. There are many aspects to the problem and hence the goal will be difficult to reach. It involves many professions responsible for building or modifying a city: Architects and engineers, health personnel and regulators, landscape architects and gardeners, construction and building enterpreneurs, and finally those who deal with the physical environment, such as hydrologists and climatologists. It is readily seen that these diverse skills and interests often operate independently and in conflict with each other. The task of the planner is to coordinate their work, arrive at an optimal design, and reconcile the conflicting desires. This is a Herculean job, especially because large economic values are at stake.

Each profession can only add a narrow facet to the total design spectrum. The climatologist has in the past, more often than not,

255

been left out of the planning process. In fact, sometimes the most primitive, if not ludicrous, reasoning about atmospheric conditions has been incorporated into urban plans. In many eastern United States cities the planners have arranged industrial zones on the eastern fringes. They had heard about the "prevailing westerly winds" and thought this location would carry the industrial air pollutants downwind from the urban area. But the prevailing (or most frequent) wind direction is not the relevant factor. Rather, the wind directions for the so-called stagnation weather patterns are pertinent. These are often easterly in that area, rather than westerly and would carry, with low mixing heights, the polluting plumes back over the city. This is only one of many examples.

The main trouble in the past has been inadequate communication between meteorologists and the other professions. People have always looked at the atmosphere and weather with a fatalistic attitude. As Oke (1976) has pointed out: "Until very recently engineers, architects and planners were largely unaware of the atmospheric implications of their activities (indeed many atmospheric scientists were also ill informed in this area). Some members of these professions are now awakened to these possibilities and they are posing practical questions as to how to predict or to minimize these effects, some of which constitute feedback loops."

Some international agencies, the World Meteorological Organization (WMO) and the Conseil International du Bâtiment (CIB) have tried to bridge some of the existing gaps. Among their literature a broad review by Chandler (1976) tries to indicate what the contribution from the climatological field for more rational urban planning might be. It is, of course, much easier to plan a town from its beginning than to redesign an existing one (Landsberg, 1973).

An instructive case comes to mind. That is the town of Kitimat in British Columbia, where an aluminum smelter was to be set up and the necessary housing and amenities for the workers were to be provided, including an air strip. It was readily predictable that after deforestation of the slopes, low-lying areas would be flooded from time to time due to the heavy snowfalls in the area. This dictated land use for that part of the town area for recreation and gardening. During inversion conditions, mountain winds were to blow smelter pollutants out to the fjord. Higher elevations were advantageous for specialized installations such as the hospital. Appropriate plantings along the highway from the town to the smelter could act as snow fences and minimize drifting (Anonymous, 1954).

In existing towns the greatest damages from weather events come from floods, the most insiduous ones from air pollution, and the most annoying ones from wind modifications. The heat island, as discussed earlier, has some benefits and some detriments. On balance it would seem to be most desirable to reduce or eliminate as many of the man-made local climate modifications as possible. Table 11.1 gives an overview of these changes.

Easiest to avoid are flood problems along streams and rivers. Channelling and reservoir control by dams will help. But more important in planning is the complete avoidance of flood plains for housing or industry. In older settlements after major floods it is far better to withdraw land from that type of use than to rehabilitate it. Some states now have laws to prevent that kind of misuse of land. It is axiomatic that flood stages once reached in what may seem to be a unique rainfall event, such as the passing of a tropical storm, will sooner or later be repeated. For tributaries running through urban areas adequate storm drainage must be provided. The common practice to use the 2-yr maximum daily rainfall value as a design factor is indefensible, especially in recognition of rainfall increases by urbanization and the plausibility of an increase in thunderstorm activity (Changnon, 1979).

Land use also seems to become important when urban-induced rainfall presents problems. Changnon (1979) presents a long list of such problems. They include, among others: bypassing of sewage treatment plants by runoff, automobile accident increases, and higher cost of water management. Auer (1978) analyzed some of the consequences of land use in St. Louis, especially where vegetation covers less than 35 percent of the ground, and that suggested that dispersal of industrial, commercial, and compact residential area into smaller zones, less than 20 km^2, would be a remedy for the increased urban rainfall.

Only marginal improvements are possible for the urban heat island. These are primarily needed for the hot, sunny summer days. Aside from provision of trees for shading and sprinkling of streets to increase evaporative cooling, the only other remedy is an increase in albedo. In hot areas this can be provided by light paint on the walls and roofs of buildings. The elimination of parking lots and their

TABLE 11.1

Climatic Alterations Produced by Cities

Element	Compared to rural environs
Contaminants:	
condensation nuclei	10 times more
particulates	10 times more
gaseous admixtures	5–25 times more
Radiation:	
total on horizontal surface	0–20% less
ultraviolet, winter	30% less
summer	5% less
sunshine duration	5–15% less
Cloudiness:	
clouds	5–10% more
fog, winter	100% more
summer	30% more
Precipitation:	
amounts	5–15% more
days with <5 mm	10% more
snowfall, inner city	5–10% less
lee of city	10% more
thunderstorms	10–15% more
Temperature:	
annual mean	0.5–3.0°C more
winter minima (average)	1–2°C more
summer maxima	1–3°C more
heating degree days	10% less
Relative humidity:	
annual mean	6% less
winter	2% less
summer	8% less
Wind speed:	
annual mean	20–30% less
extreme gusts	10–20% less
calm	5–20% more

replacement by subsurface parking or parking garages would be helpful, because the combination of asphalt lots and cars acting as miniature greenhouses not only leads to very high surface temperatures but also to the unbearable heat in cars.

There is universal agreement on the use of green surfaces to improve urban climates (Laurie, 1979). In summer they increase cooling by evapotranspiration. They reduce rapid runoff and, if judiciously planted, will reduce particulate concentrations. They generally increase urban albedos (Bach and Mathews, 1969; Böhm, 1979). Certainly strips of plants will reduce noise and pollution. In some cities, Volgagrad is a notable example, sizeable green belts separate industrial from residential sectors. Very little quantitative information has become available on such experiments, but data from older cities such as Vienna indicate beneficial effects. Modern cities with their evermore intense land utilization have, in many places almost completely eliminated vegetation. Lorenz (1973) in his flight measurements of surface temperatures pointed out that roofs show among the highest values, in contrast to grass surfaces. There is no serious experimentation with parklike roof gardens. But green areas in cities have to reach a vertical size and dispersal to become effective as a suitable measure for improvement of urban climate.

Wind problems now plague many inner cities. The urban canyons (Oke, 1976) channel the air flow and at times create intense Venturi effects. Tall buildings produce highly undesirable eddies. Some of these adverse conditions can be avoided by adequate street width and building design. Occasionally street orientation can mitigate difficulties, especially where specific wind directions and speeds can be expected to create adverse conditions. There is some evidence that uniform building height and uniform distances between buildings create the least flow disturbances (Wilmers, 1975).

In many localities it is essential to adapt the urban design to existing natural flow patterns. Thus near water bodies the daily lake or sea breeze may be a most welcome relief from excessive heat. Blocking this breeze by a solid barricade of tall buildings at the water front will deprive areas further inland of the beneficial effects. Similarly judicious planning can, in hilly terrain, take advantage of noctural mountain breezes to help in dispersal of air pollutants.

Air pollution remains, of course, the most difficult urban atmospheric problem to counteract. Wherever feasible, eliminating pollutants at the source is most desirable. The economic and engineering obstacles to such a course of action require us to look for alternatives. The greatest, most ubiquitous source of urban air pollution is the motor vehicle, powered by an internal combustion engine using the common gasoline or diesel fuels. A combination of provision for alternate transportation is the essential task for the city planner. Electric transit, electric cars or cars fueled by hydrogen is one of the solutions. Neighborhoods where all necessary facilities are within walking or cycling distance will help. Providing space heating by electricity or natural gas is another remedy. However, the use of electricity, unless produced by water, solar, or geothermal power will only transfer the production of pollutants elsewhere. This may be sufficiently far from the city to help the urban problem.

For industrial installations in urban areas an ancient proposal may have to be revived, namely the smoke sewer. This was first proposed over 100 yr ago. In West Germany this scheme has been applied in some locations to duct smelter fumes from valley towns to nearby heights where they were discharged in an environment with good atmospheric dispersal.

In cities where snow is a common occurrence great benefit would accrue from the use of waste heat under streets and sidewalks. In all cities in the higher latitudes substantial amounts of heat are discharged in winter from chimneys and stacks. Were this heat channeled to the roads, it could be eminently beneficial. A minimal effort would at least protect criticial spots in the traffic system. Among these are bridges, slopes, and crossings. Particularly dangerous are glaze storms. In those cases supercooled water falls on a surface with temperatures below freezing point and freezes on impact. The result is a glaring ice surface. Such conditions occur in the snowbelt of the United States once every 2 or 3 yr. Snowfalls of 20 cm or more are quite frequent. Such conditions can paralyze a city. Offices and schools are closed, transportation systems are impeded, and utility failures occur. Climatic analyses have shown that planning should be for snowfall rates of 2.5–7.5 cm hr^{-1}. Melting 1 cm of ice requires theoretically 142 W m^{-2}, but in practice twice that amount is needed because part of the energy in ducts under a road flows downward. Installation costs are lowest if the work is done during construction of a road or when repaving is done. If waste heat is used operating costs are minimal. Over a period of time the costs are small compared to the investments in snow plows and sanding trucks, the cost of operation, and the cost of sand and salt. The savings in the reduction of accidents in medical costs for injuries and vehicle damage are substantial (Coulter and Herman, 1965; Michalski, 1965).

The energy-efficient city of the future must be planned in harmony with what climate has to offer. It must be designed to meet the climatic hazards and to mitigate the well-known climatic alterations of urbanization (Landsberg, 1970).

References

Anonymous (1954). Kitimat: A new city. *Archit. Forum* (7), 128–147; (8) 120–127; (10) 158–161.

Auer, A. H. (1978). Correlation of land use and cover with meteorological anomalies. *J. Appl. Meteorol.* **17**, 636–643.

Bach, W., and Mathews, E. (1969). The importance of green areas in urban planning. Paper presented at workshops, *Bioclimatol. Environ. Health, Cincinnati, July 14–16, 1969*, 20 pp. U. S. Dept. Health, Education, and Welfare, Public Health Service, Washington, D. C.

Böhm, R. (1979). Meteorologie und Stadtplanung in Wien- ein Überblick, *Wetter Leben* **31**, 1–11.

Chandler, T. J. (1976). Urban climatology and its relevance to urban design. *WMO Tech. Note,* No. 149, 61 pp.

Changnon, S. A. Jr. (1979). What to do about urban-generated weather and climate changes. *APA J.* (Jan. 1979), 36–47.

Coulter, R. G., and Herman, S. (1965). Snow and ice control by induced melting. *In* "Snow Removal and Ice Control in Urban Areas" (R. K. Lockwood, ed.), Special Rept. No. 30, pp. 11–14. Am. Publ. Works Assoc., Washington, D.C.

Landsberg, H. E. (1970). Climates and urban planning. *In* "Urban Climates," *WMO Tech. Note,* No. 108.

Landsberg, H. (1973). The meteorologically utopian city. *Bull. Am. Meteorol. Soc.* **54**, 86–89, 364–372.

Laurie, Ian C., ed. (1979). "Nature in Cities—The Natural Environment in the Design and Development of Urban Green Space," 428 pp. Wiley, New York.

Lorenz, D. (1973). Meteorologische Probleme bei der Stadtplanung. *Forshungsgemeinschaft Bauen und Wohnen FBW—Blätter, Baupraxis* 9/73, 57–62.

Michalski, C. S. (1965). Effect of winter weather on traffic accidents. *In* "Snow Removal and Ice Control in Urban Areas" (R. K. Lockwood, ed.), Special Rept. No. 30, pp. 119–122. Am. Publ. Works Assoc.

Oke, T. R. (1976). Inadvertent modification of the city atmosphere and the prospects for planned urban climates. *Proc. WMO Symp. on Meteorology as Related to Urban and Regional Land-Use Planning, WMO,* No. 444, 150–175.

Wilmers, F. (1975). Klimatologische Überlegungen zur Bebauung von Freiflächen in Stadtgebieten besonders am Rande von Flussniederungen. *N. Arch. Niedersachsen* **24**, 259–277.

Author Index

Numbers in italics refer to the pages on which the complete references are listed.

A

Ackerman, B., 2, *14*, 143, 145, *149*
Ackerman, T. P., 61, *79*
Agee, E. M., 205, *207*
Aichele, H., *22*
Aida, M., 61, 62, *79*
Allee, P. A., 205, *209*
Anderson, D. G., 213, 215, 218, 219, *222*
Anderson, S. F., 148, *150*
Angell, J. K., 142, 145, 146, 148, *150, 151*
Arakawa, H., 136, *150*
Arens, E., 248, *251*
Ariel, M. Z., 138, *150*
Ashby, W. C., 195, *207*
Ashworth, J. R., 193, *207*
Aston, A. R., 76, 77, *80*
Atkinson, B. W., 198, 199, 200, *207*
Atwater, M. A., 62, *79*, 109, *123*, 161, 163, *174*
Auer, A. H., Jr., 60, *82*, 143, *150*, 257, *261*
Auliciems, A., 147, *150*
Averett, R. C., 222, *222*

B

Bach, W., 76, 79, *79*, 155, *174*, 232, *251, 252*, 259, *261*
Badger, N. K., 134, *150*
Ballanti, D., 248, *251*
Ballestin, H., 227, *251*
Barad, M. L., 135, *151*
Barrett, E. W., 193, *209*
Bauman, S., 227, *251*
Beilke, S., 43, *50*
Belger, W., 148, *150*
Bender, G. M., 216, 217, 218, *222*
Benson, F. B., 242, *252*
Berg, H., 134, *150*
Bergstrom, R. W., 40, 41, *50*, 62, *82*, 171, 172, *174, 175*
Berkes, Z., 195, *207*
Bhumralkar, C. M., 156, 160, *174*
Bielich, F.-H., 56, *80*
Binkowski, F. S., 138, *151*
Birenfeld, C., 248, *254*
Bishop, D. E., 228, *252*
Black, C. H., 93, 96, *124*

263

Subject Index

The term *urban* should be understood to precede most of the main entries.

A

Acid rainfall, *see* Rainfall
Aerosol, 23, 186, 192, 193
Air pollution
  concentration, 169–172
  damages, 223–226, 257
  emission, 169, 170
  episodes, 49, 236, 237
  health effects, 29, 236–243
  history, 3
  index, 29
  observations, 20
Air quality
  observations, 27
  standards, 26, 27, 29
Aitken nuclei, *see* Condensation nuclei
Albedo, 59–61
Anthropogenic influence on rainfall, *see*
  Precipitation
Arteriosclerotic heart disease, 238
Asthma, 238
Atmosphere, 23–26
  chemistry, 24–26

gases, 23
urban, 23–49
Atmospheric interaction, synoptic–
  local, 17–18
Atmospheric radiation, *see* Radiation

B

Bronchitis, 238

C

Cadmium, 238, 240
Carbon monoxide, 27, 30, 33, 45, 241–
  243
Carboxyhemoglobin, 241–242
City fog, 4
Civil disorders, 250, 251
Climate model, *see also* Models
  general, 11
  physical, 12
Climatic alterations, 258

International Geophysics Series

EDITED BY

J. VAN MIEGHEM
(July 1959–July 1976)

ANTON L. HALES
(January 1972–December 1979)

WILLIAM L. DONN
Lamont-Doherty Geological Observatory
Columbia University
Palisades, New York